YALE AGRARIAN STUDIES SERIES

James C. Scott, series editor

For a complete list of titles in the Yale Agrarian Studies Series, visit yalebooks.com/agrarian.

CORRIDORS
OF
POWER

The Politics of Environmental
Aid to Madagascar

CATHERINE A. CORSON

Yale
UNIVERSITY PRESS
New Haven and London

Published with assistance from the foundation established in memory of Philip Hamilton McMillan of the Class of 1894, Yale College.

Yale University Press books may be purchased in quantity for educational, business, or promotional use. For information, please e-mail sales.press@yale.edu (U.S. office) or sales@yaleup.co.uk (U.K. office).

Set in Janson type by IDS Infotech, Ltd.
Printed in the United States of America.

Library of Congress Control Number: 2016930341
ISBN 978-0-300-21227-3 (cloth : alk. paper)

A catalogue record for this book is available from the British Library.

This paper meets the requirements of ANSI/NISO Z39.48-1992 (Permanence of Paper).

10 9 8 7 6 5 4 3 2 1

In Memory of
Alison Jolly (1938–2014)

In appreciation for sharing her insights into the passionate history of conservation politics in Madagascar. I offer this book in acknowledgment of the debates we might have had.

Contents

Preface

When you drive around Madagascar, you see the land burning. You see fires burning all over the country. People are burning down the forest for farmland. It is a type of slash and burn agriculture, and it is the biggest threat to lemurs in Madagascar.

—*ISLAND OF LEMURS: MADAGASCAR,* Imax movie

Do you have any "live" people? No, only dead ones. I mean, if we had a bunch of live people running around, it wouldn't be called "the wild," would it?

—INTERCHANGE BETWEEN THE CHARACTERS ALEX, MAURICE, AND KING JULIEN XIII in *Madagascar the Movie*

I REMEMBER WELL MY first flight to Madagascar. I was working for the U.S. Agency for International Development's Africa policy office at the time and visiting USAID missions in East Africa. As we flew from Nairobi, Kenya, toward Antananarivo, or "Tana," the capital of Madagascar, I looked down, expecting to see stretches of rainforest. Instead, to my surprise, I saw a brown, flat, barren landscape—the highlands plateau. It was early November, and, as I stepped off the plane in Tana, the simultaneously sweet and bitter smell of the grassland fires filled the air. It was a reminder that Madagascar is not "the wild," but, like any country, a complex landscape, both shaped by and shaping human cultures and political economic systems over centuries.

Every year a quarter to a half of Madagascar's grasslands and pastures are burned, as Christian Kull documents in his book *Isle of Fire*. Although the vast majority of this burning is not in forests, the striking image of the countryside in flames has contributed to the narrative that Madagascar was once completely forested and that 80–90 percent of its forest has been destroyed by growing populations of slash-and-burn agriculturalists who continue to destroy hundreds of thousands of hectares of pristine forest each year. The solutions, this narrative argues, are to educate these farmers about the destructive nature of their inefficient farming techniques, to cordon off areas of high biodiversity from human use, and to create economic incentives to motivate farmers to conserve biodiversity.

Numerous scholars have challenged this hegemonic narrative, arguing that the island was never completely forested and that there are a number of other, interrelated and regionally specific drivers of deforestation, such as timber exploitation, mining, and large-scale agriculture. These critics underscore the risks of analyzing Madagascar's environmental challenges in geographic isolation and the importance of understanding the history of its human-environment dynamics. They stress that rural peasants have faced a long history of political and economic marginalization, which shapes how they interact with the landscape. Many live on marginal lands with limited options, and they often clear land not only for crops but also in order to claim it *back* from either the state, conservationists, commercial extractors, or migrants who have expropriated it. They also emphasize that the ways in which people use land and resources are intertwined with global patterns of consumption, trade, and foreign aid as well as with perceptions of nature and people. Importantly, these critics are arguing that the narrative blaming forest fragmentation on shifting cultivators is incomplete, not that shifting cultivation is inconsequential.

Why is this incomplete explanation so pervasive despite complicating evidence? First, if one arrives in Madagascar with the preconception that it is or should be covered in rainforest, this narrative is easy to believe. Along the main roads to the regional capitals of Fianarantsoa and Toamasina—the roads most traveled by westerners—one can see patchworks of burned or burning hillsides as well as *lavakas,* where the soil has eroded away from the denuded hills. These cultivated landscapes contrast sharply with Western ideas about pristine environments. Furthermore, viewing landscapes at a particular moment in time or even over several decades conceals the history of peasant–state conflicts over forests. Likewise, we tend to see landscapes as bounded, ignoring the extralocal forces, such as the increasing global demand for Madagascar's natural resources, which affect local land use decisions. Finally, even as many consultants, government officials, donors, conservationists, researchers, and villagers strive to tell this more complicated story—and to redirect policies, programs, and funding accordingly—it is harder to "sell" in the concise sound bites that make movies and attract foreign aid funding. They strategically invoke narratives that will attract political, financial, and media support because they resonate in policy circles and are accepted as truths, even as they know them to be partial.

Enduring narratives tend to reflect and reinforce the interests of powerful actors. They persist even in the face of complicating evidence precisely because they benefit those who perpetuate them. Environmental degradation narratives often blame those with the least ability to voice their concerns and avoid mentioning the impacts of the more powerful. The narrative that attributes Madagascar's deforestation primarily to rural peasants while eliding other critical drivers of deforestation has long legitimized efforts by the state to control the country's forests and to expropriate rural peasants' land for large-scale commercial enterprises. As McConnell and Kull argue, the assertion that Madagascar was once completely forested has formed a "motivating *raison d'être*" for strong, foreign-funded conservation interests to intervene in the island's resource politics. The voices of those who have the biggest stake in telling the more complex story of deforestation in Madagascar—the rural peasants—are harder to find in the corridors of power.

Ultimately, all narratives, including this one, are incomplete, designed to bind the storyteller and listener in a relationship. They are shaped by the subjective knowledge and perspective of both parties, and they exaggerate aspects that will draw listeners in and leave out details that will alienate them. The narratives that resonate at particular points in time change as they shape and are shaped by political, economic, social, and cultural contexts. Nonetheless, as we hear explanations of environmental change repeated over and over again, we forget that they represent conscious political strategies to build relationships. They begin to influence what we see in a landscape, whom we blame for its condition, and what we think should be done about it. In doing so, they affect who gets to use resources for livelihoods or profit and who controls, manages, and benefits from them. Thus to understand the drivers of degradation and efforts to redress it, we must ask, "What political relationships do current explanations and solutions reinforce and how?"

I argue that the rise of U.S. funding for biodiversity conservation in Madagascar emerged out of the combined consolidation of state authority within global environmental institutions and neoliberal reforms, which brought nonprofit and private actors into positions of authority in state policy-making processes. The resulting assemblages of public, private, and nonprofit actors could align behind the framing of Madagascar's environmental issues as a peasant problem and endorse biodiversity conservation through the expansion of protected areas as a politically feasible way to be green. While this was an effective strategy to mobilize critical political and financial support, it elided the complexity of drivers of environmental degradation in Madagascar. The resulting apportionment of programs marginalized small-scale rural peasants, failed to invest sufficiently in community engagement and state capacity-building needed to ensure effective long-term conservation, and left the country vulnerable to increasing pressure to overexploit its natural resources.

In writing this book, I join the collective endeavor by consultants, policy makers, and scholars who try to explain how and why environmental degradation in Madagascar worsens despite the millions invested in "saving" the environment there. Using USAID's environmental program there as a window into the political economy of conservation, I illustrate how neoliberal reforms and the formation of global environmental institutions changed who had the authority to shape environmental degradation narratives, policy priorities, and programmatic strategies.

As my narrative moves from the corridors of Congress to villages in eastern Madagascar, I draw on ethnographic research conducted across multiple countries, organizations, and events to illustrate how various actors conceive of, develop, and implement U.S. environmental foreign aid. I use observations, interviews, and archival research to show how these actors come together and align at particular nodes and, in doing so, create a transnational, dynamic field of governance.

My commitment to using ethnography to study policy-making processes is explicitly grounded in a decade of professional work in environment and development policy, research, and consulting, and many of the transformations in global environmental politics I discuss in this book happened within my lifetime. Like many political ecologists, I began my studies in biology. I first heard Alison Jolly speak about lemurs as a neurobiology and behavior major at Cornell University, and I worked as a research assistant at the Caribbean Primate Research Center and Laboratory of Ornithology. Intrigued by the rising popularity in the late 1980s and early 1990s of integrated conservation and development projects, I went to Zimbabwe after graduating to study participatory development. From there I moved on to Washington, D.C., in the mid-1990s, in time for Jesse Helms's foreign aid reforms and Al Gore's government downsizing project that aimed to "Do More with Less." From vantage points in the White House, the State Department, and the National Oceanic and Atmospheric Administration, I sat in meeting rooms, coffee shops, and hotel lobbies collaborating with environmental nongovernmental organization representatives to develop political strategies and to draft policy and legislation. In the Senate Appropriations Committee I kept track of congressional directives and worked with representatives of both private and nonprofit organizations to craft report and bill language. Finally, as an economic advisor in USAID's Africa Bureau, I worked to implement the shift from state-led to market-oriented economic development policies, encompassed in what has been termed the Washington Consensus and its subsequent embrace of poverty reduction in a post–Washington Consensus. Through these experiences I came to appreciate the dialectic between the everyday actions of individuals, their experience of bureaucratic constraints, and the broader global political economic transformation that has happened over the past forty years. I hope to give academics and aid recipients alike an enhanced understanding of what shapes the ways in which foreign aid policy makers define, bound, and try to address environmental problems and in the process to reveal multiple avenues for change.

With these aims in mind, I dedicate this book to the late Alison Jolly, who first began working in Madagascar in the 1960s and who shaped conservation politics there for many decades. Even as she knew I was, in her words, a " 'strong critic' of western conservation imperialism," she opened up her archival material to me and shared her personal experiences. I spent many days in her back room reading her notes and diaries about the efforts of a small group of people who in the 1970s sought to raise awareness of the plight of Madagascar's flora and fauna. Alison probably would have disagreed with much of what I argue here. Nonetheless, I offer it as recognition of our joint interest in catalyzing more effective, but also socially just, conservation. While many who read this book will view it as anticonservationist, its intent is the opposite, and its critique should not be taken as a lack of respect for those who have worked so hard over the course of their lives to save Madagascar's biodiversity.

By documenting the processes through which environmental aid is pursued, I hope to underscore that because policy narratives and associated practices reflect particular relations of power, we cannot redress misguided policies simply by investing in community consultation or state capacity building. Ultimately we must transform the transnational power relationships that have created these policies and that are sustained by them. This will demand changing who has authority to shape conservation politics—that is, who travels through the corridors of power.

Acknowledgments

I AM GRATEFUL TO the countless people who went out of their way to facilitate this research. A number of people provided feedback throughout the initial research and writing. At the University of California, Berkeley, I am tremendously indebted to Louise Fortmann, Nancy Peluso, Isha Ray, and Gillian Hart as well as to Allan Hoben, George Scharffenberger, Christian Kull, Elizabeth Shapiro, Daniel Graham, and in particular Nathaniel Gerhart, who passed away while doing fieldwork in Indonesia. His passion, commitment, humor, and dedication to his friends and family were a gift to us all. Innumerable times while writing this book I wished I could have asked his advice. Above all, Jennifer Casolo generously shared some of her brilliance on countless occasions to help me find the deeper theoretical implications of my work. At Cambridge University, Bill Adams, Sarah Milne, Ivan Scales, Rob Small, and especially Jennifer Talbot were supportive sounding boards. At the University of Sussex, James Fairhead, Buzz Harrison, Melissa Leach, and Ian Scoones helped me develop the book's theoretical contributions, and at Mount Holyoke College, Liz Garland provided valuable feedback, and Eugenio Marcano assisted with the finishing touches on the maps.

Throughout my research the staff at the USAID-Madagascar mission was enormously helpful, including Henderson Patrick, Daniela Raik, Zoelimalala Ramanase, Benjamin Rakotondranisa, Eddy Rasoanaivo, Johanesa Rasolofonirina, Tiana Razafimahatratra,

Josoa Razafindretsa, Ndrantomahefa Razakamanarina, Jennifer Talbot, Viviane Voahanginiriana, and especially Lisa Gaylord. Lisa ran the USAID-Madagascar environment program for many years and continues to be an influential figure in Madagascar conservation politics. I am grateful for her willingness to share her plethora of knowledge and experience as a USAID environmental official in Madagascar and her review of chapter 6. From her I learned a tremendous amount, and I have great respect for her tireless commitment and the continual efforts she made to improve not just USAID's program but also the Madagascar environmental program in general.

I also want to thank Alison Jolly, Russ Mittermeier, and Alison Richard, all of whom reviewed drafts of chapter 3. Likewise, Jean-Paul Paddock of WWF-Madagascar, and Tom Erdmann, Mark Freudenberger, and Vololona Raharinomenjanahary of the Ecoregional Initiatives program went above and beyond the call of duty to share their time and reflections and to offer general assistance. Olga Ranivoriaka at the Bibliothèque Grandidier in Antananarivo was indispensable in helping me sort through archival material. Similarly, the staff at the USAID Development Experience Clearinghouse and the Library of Congress helped me locate USAID and congressional documents. Finally, I am beholden to all of the people who took time out of their busy schedules to meet with me and who were willing to be interviewed during this research. My conclusions are my own and do not reflect USAID's official position or the opinions of USAID staff or any other interviewees.

Most important, I am deeply grateful to my research collaborators and assistants in Madagascar and the United States. Mijasoa Andriamarovololona and others at Vokatry ny Ala facilitated my field visits and were a constant source of grounding about conservation in Madagascar. Philippison Andrianjasoa Lee, Gloria Moroyandsa, Lalanirina Andrianisa, Aina Rakotondrazaka, Vivia Moroyandsa, Feno Moroyandsa, Ainariravako Andriantseheno, Mahareta Paubert, Hafina Raharinomenjanahary, Sahoby Randriamahaleo, and Ando Ratovona all assisted with research, transcription, and translation. Most of all, I would like to thank Salohy Ratovona, who was a fabulous research assistant before leaving to undertake her own doctoral degree. Last, I am grateful to the Malagasy families who welcomed us into their lives. In line with my institutional review board agreement, I have chosen to keep the names of interviewees and the villages I visited confidential. I am indebted to undergraduate students at the University of California at Berkeley and Georgetown University who assisted with research, transcription, translation, editing, and maps, including Caitlin Hachmyer, Sylvia Ewald, and Kim Howell as well as Julia Worcester, Hannah Kyer, Katy Johnston, Maeve Moller-Mullen, Kate Rawson, and Amelia Neumayer at Mount Holyoke College.

I would like to thank three anonymous reviewers of my prospectus and book as well as the reviewers of related articles for their helpful feedback and guidance along the way. And I offer special thanks to Samantha Ostrowski, Jeffrey Schier, and Lawrence Kenney for their assistance in the preparation of the final manuscript and to my editor at Yale University Press, Jean Thomson Black, not only for taking a chance on my book but also for her patience and support over the years. My wonderful colleagues at Mount Holyoke, Lauret Savoy, Tim Farnham, Kate Ballantine, and Donna McKeever, helped protect my time as I finished the book, and my co-conspirators in collaborative ethnography Lisa Campbell, Noella Gray, Ken MacDonald, Peter Wilshusen, Pete Brosius, Rebecca Gruby, Kim Marion Suiseeya, Dan Suarez, Sarah Milne, and others helped me think through the value of using ethnography to study policy-making processes.

Finally, I would like to express immense gratitude to my husband, Desmond, who accompanied me to Madagascar and, even as he was writing his own book, read innumerable versions of mine and helped to take care of our two young boys while I finished writing . . . and finished again . . . and again. And to my parents, David and Carolyn, who stepped in to provide childcare on several occasions. Last but not least I acknowledge Geoffrey and Alex, who came into this world as I finished my dissertation and who have been growing up as I have written this book. Their presence reminds me daily how important it is not to give up trying to make the world a better place for their generation.

Research and writing support came from the National Science Foundation; American Association of University Women; Andrew W. Mellon / American Council of Learned Societies; Woodrow Wilson

International Center for Scholars; Rural Sociological Society; Foreign Language and Area Studies Program; Frank Myers Forestry Scholarship Program; Center for African Studies at the University of California, Berkeley; Beahrs Environmental Leadership Program; Mount Holyoke College Faculty Grants; and graduate student research and teaching support from the Department of Environmental Science, Policy, and Management at the University of California, Berkeley, Mount Holyoke College, the Department of Geography at Cambridge University, the Department of Anthropology at the University of Sussex, the Woodrow Wilson Center for International Scholars, and the Washington, D.C., office of the University of California all provided office space and other technical support for the research and writing.

Portions of chapter 4 have been published previously in *Antipode*, http://onlinelibrary.wiley.com/doi/10.1111/j.1467-8330.2010.00764.x/full, and portions of chapter 7 have been previously published in *Society and Natural Resources*, http://www.tandfonline.com/doi/abs/10.1080/08941920.2011.565454, *The Journal of Peasant Studies*, http://www.tandfonline.com/doi/full/10.1080/03066150.2011.607696, and in a chapter of *Conservation and Environmental Management in Madagascar*, edited by Ivan Scales and published by Earthscan.

Abbreviations

AGERAS	Support to Regional Environmental Management and the Landscape Approach (Appui à la Gestion Régionalisée de l'Environnement et à l'Approche Spatiale)
AGEX	Executing Agencies (Agences d'Exécution)
ANAE	National Association for Environmental Actions (Association Nationale d'Actions Environnementales)
ANGAP	National Association for the Management of Protected Areas (Association Nationale pour la Gestion des Aires Protégées)
ARD	Associates in Rural Development, Inc.
BATS	Biodiversity Analysis and Technical Support
BSP	Biodiversity Support Program
CAPAE	Support Center for the Environmental Action Plan (Cellule d'Appui au Programme d'Action Environnementale)
CARE	Cooperative for Assistance and Relief Everywhere
CBD	Convention on Biological Diversity
CBNRM	Community-Based Natural Resource Management
CDO	cooperation and development organization

CI	Conservation International
CIRAD	Center for International Cooperation in Agronomic Research for Development (Centre de Coopération Internationale en Recherche Agronomique pour le Développement)
CNCD	National Commission for Conservation and Development (Commission Nationale de la Conservation pour le Développement)
COAP	Code for Managing Protected Areas (Code de Gestion des Aires Protégées)
COBA	Natural resource management community association (Communauté de Base)
COMODE	Malagasy Council of NGOs for Environment and Development (Conseil Malgache des Organisations Non-Gouvernementales pour le Développement et l'Environnement)
CoP	Conference of the Parties
CRS	Congressional Research Service
CSD	Commission on Sustainable Development
CTP	Permanent Technical Committee (Comité Technique Permanent)
DCT	Development and Conservation Territory (Territoire de Conservation et Développement)
DGEF	General Directorate of Water and Forests (Direction Générale des Eaux et Forêts)
EDF	Environmental Defense Fund
EIA	Environmental Investigation Agency
EP1	Environmental Program 1
EP2	Environmental Program 2
EP3	Environmental Program 3
ERI	Ecoregional Initiatives
FAA	Foreign Assistance Act
FAO	United Nations Food and Agricultural Organization
FDI	foreign direct investment

FORAGE	Regional Funds to Support Environmental Management (Fonds Régional d'Appui à la Gestion de l'Environnement)
FY	fiscal year
GAO	Government Accountability Office
GCF	Contract-based Forest Management (Gestion Contractualisée des Forêts)
GDP	gross domestic product
GEF	Global Environment Facility
GELOSE	Secure Local Resource Management (Gestion Locale Sécurisée)
GIS	geographic information systems
GTC	Global Tomorrow Coalition
GTZ	German Cooperation Agency (Deutsche Gesellschaft für Technische Zusammenarbeit)
HIPC	Highly Indebted Poor Countries
IAGS	International Advisory Group of Scientists
ICB	International Conservation Budget
ICC	International Conservation Caucus
ICCF	International Conservation Caucus Foundation
ICDP	Integrated Conservation and Development Project
ICP	International Conservation Partnership
IDCA	International Development Cooperation Agency
IFI	International Financial Institutions
IMF	International Monetary Fund
IRD	Research Institute for Development (Institut de Recherche pour le Développement)
IRG	International Resources Group
IRSM	Institute of Scientific Research in Madagascar (Institut de Recherche Scientifique de Madagascar)
IUCN	World Conservation Union (Union Internationale pour la Conservation de

	la Nature et de ses Ressources), formerly International Union for the Conservation of Nature
JWPT	Jersey Wildlife Preservation Trust
KEPEM	Knowledge and Effective Policies for Environmental Management Project
KfW	German Development Bank (Kreditanstalt für Wiederaufbau)
LDI	Landscape Development Interventions
MAP	Madagascar Action Plan
MBG	Missouri Botanical Garden
MCA	Millennium Challenge Account
MCC	Millennium Challenge Corporation
MDG	Multi-Donor Group
MDS	Multi-Donor Secretariat
MECIE	Ensuring Investment Compatibility with the Environment (Mise en Compatibilité des Investissements avec l'Environnement)
MESupReS	Ministry of Higher Education and Scientific Research (Ministère de l'Enseignement Supérieur et de la Recherche Scientifique)
MFG	Madagascar Fauna Group
MinEnvEF	Ministry of Environment, Water, and Forests (Ministère de l'Environnement, Eaux et Forêts)
MRSTD	Ministry of Scientific Research and Technology for Development (Ministère de la Recherche Scientifique et Technologique pour le Développement)
NEAP	National Environmental Action Plan
NEPA	National Environmental Policy Act
NGO	nongovernmental organization
NHWP	Nature, Health, Wealth, and Power
NRDC	Natural Resources Defense Council
NWP	Nature, Wealth, and Power
NYZS	New York Zoological Society

ONE	National Environment Office (Office National pour l'Environnement)
OSF	Forest Observatory (Observatoire du Secteur Forestier)
PACT	Private Agencies Collaborating Together
PAGE	Environmental Management Support Project (Projet d'Appui à la Gestion de l'Environnement)
PRSP	Poverty Reduction Strategy Paper
PVO	private voluntary organization
QMM	QIT Madagascar Minerals
REBIOMA	Madagascar Biodiversity Network (Réseau de la Biodiversité de Madagascar)
REDD+	Reducing Emissions from Deforestation and Forest Degradation in Developing Countries
RFA	request for applications
RFP	request for proposals
SAGE	Environment Management Support Service (Service d'Appui à la Gestion de l'Environnement)
SAPM	System of Protected Areas in Madagascar (Système d'Aires Protégées de Madagascar)
SAVEM	Sustainable Approaches to Viable Environmental Management
SMB	Multi-Donor Secretariat (Secrétariat Multi-Bailleurs)
SSC	Species Survival Commission
TNC	The Nature Conservancy
TRD	Tropical Research and Development
UN	United Nations
UNDP	United Nations Development Programme
UNESCO	United Nations Educational, Scientific, and Cultural Organization
USAID	United States Agency for International Development
USDA	United States Department of Agriculture

WCS	Wildlife Conservation Society
WPC	World Parks Congress
WRI	World Resources Institute
WWF-International	World Wide Fund for Nature
WWF-Madagascar	World Wide Fund for Nature Madagascar
WWF-U.S.	World Wildlife Fund U.S.

CORRIDORS OF POWER

CHAPTER ONE

Connecting Corridors

You understand, in Madagascar we must succeed. We haven't
a choice. It is not like Europe here, or America. If you
desertify part of your country, you have enough to spare. . . .
If we destroy Madagascar and turn our island into a desert,
what will we do? We'd have to swim!

—GUY RAZAFINDRALAMBO, November 12, 1989, quoted in
Jolly 2015, 143

It is easier to do biodiversity overseas than in the United
States because the conflicts don't involve constituencies of
Congress. When there are problems with local communities,
they don't call up their congressman.

—FORMER USAID OFFICIAL, August 3, 2005

When we listen to the radio, we see different kinds of funds
from abroad in order to develop peasants in the small villages.

There is a lot of money for us from the international [community], but we can't see it here in the village. It is all just disappearing. If I spoke English, I would have gone directly to USAID and told them about our plan and our problems. But even if we say something, other people translate it the wrong way. We just see a lot of 4x4s everywhere.

—VILLAGER, MADAGASCAR, August 12, 2006

AT THE FIFTH WORLD Conservation Union (IUCN) World Parks Congress (WPC) in September 2003 in Durban, South Africa, Madagascar's former president Marc Ravalomanana announced his intention to triple the size of Madagascar's parks within five years, bringing the total surface area under protected status to 6 million hectares, or approximately 10 percent of the country's territory. The initiative—termed the Système d'Aires Protégées de Madagascar (SAPM) (System of Protected Areas in Madagascar)—aimed to meet the IUCN target of protecting 10 percent of every biome. Ravalomanana's proclamation represented a significant conservation success in one of the world's highest priority biodiversity regions. It was, in the words of the former Conservation International (CI) president Russ Mittermeier, "one of the most important announcements in the history of biodiversity conservation" (CI 2003), and it endeared Ravalomanana to conservationists for his efforts to save a country that Prince Philip of Great Britain had proclaimed, almost twenty years earlier, was "committing environmental suicide" (Jolly 2004, 210; Kull 1996, 61). By December 2010, despite the political crisis of 2009 that ousted Ravalomanana from power, there were 8.7 million hectares of protected areas and sustainable forest management sites, with an additional 10.5 million more identified as "potential sites" (Repoblikan'i Madagasikara 2010b).[1] Although the total area already exceeded the original six-million-hectare goal, at the Sixth WPC in 2014 Ravalomanana's eventual successor, President Hery Rajaonarimampianina, agreed to continue the expansion by tripling Madagascar's marine protected areas by 2020.

Two years after the announcement of 2003, as the government was in the early stages of establishing these new protected areas, I visited a U.S. Agency for International Development (USAID)–funded conservation project. This Community-Based Natural Resource Management (CBNRM) project in the eastern rainforest corridor had been initiated during the second phase of the donor- and government-coordinated Madagascar National Environmental Action Plan (NEAP) as part of an effort to devolve forest management responsibility to communities. During my visit village leaders got into a heated discussion about their increasing inability to control their land. Miners had begun prospecting on it, and a conservation organization had requested part of their farming land for the new protected area network. The villagers angrily complained that they had no choice but to relinquish their land, as the request had come from Madagascar's president; yet they knew that the remaining area was not enough to sustain a livelihood. Although compensation had not been officially offered, one villager angrily exclaimed, "Even 100 million ariary is not enough for one hectare because we earn our living from this land forever. I would not be here if my ancestors had done that."[2] It was not money he wanted but control over his means of livelihood.

A few weeks after this village meeting, the newly established International Conservation Caucus Foundation (ICCF)—a nongovernmental organization (NGO)–business partnership set up to persuade members of the U.S. Congress to fund international conservation programs, including those of USAID—held its inaugural gala in Washington, D.C. It was sponsored by four U.S.-based conservation NGOs: CI, World Wildlife Fund (WWF-U.S.), Wildlife Conservation Society (WCS), and The Nature Conservancy (TNC) as well as by transnational corporations such as International Paper, Walmart, and Exxon Mobil. The event honored the actor Harrison Ford, a member of CI's board of directors, for his commitment to conservation. In order to attend the event, guests had to pay between US$1,000 and US$50,000, the equivalent of between 2 million and 100 million ariary—the colossal amount that had been insufficient to purchase a Malagasy farmer's land.

These three events—an international conference, a village meeting, and a celebrity gala—occurred in different countries and among people of very different financial means. Yet they were deeply intertwined, connected through historical, political, and intellectual corridors of power that stretch across time and space. The ways in which people, ideas, narratives, resources, and money have come together in these corridors have been conditioned by and contributed to a forty-year reconfiguration in relations of power and authority in environmental governance under neoliberalism.

Connecting a Malagasy Village and Capitol Hill

USAID's environmental program in Madagascar offers a case study of this transformation and its relationship to shifting resource rights and access in the global South.[3] I examine the multiple historical and contemporary corridors that connect a village in eastern Madagascar to Capitol Hill in Washington, D.C. From several vantage points I analyze how various actors—representing branches of the U.S. and Madagascar governments, other donors, contractors, NGOs, scientific organizations, and villages—have developed, negotiated, and implemented U.S. environmental foreign aid to Madagascar.[4] Each chapter opens a different window into the ways in which decisions are made about Madagascar's future and different actors stake claims to Madagascar's "environment" for various purposes. Beginning in precolonial Madagascar, I explore how struggles among scientists, commercial traders, the Merina monarchy, and rural peasants over Madagascar's resources established the basis for contemporary forest politics. I then examine how a group of scientists mobilized international funding to protect Madagascar's flora and fauna. I consider how their scientifically grounded movement intersected with a U.S.-based NGO campaign to convince the U.S. Congress to earmark USAID funds for biodiversity conservation—an endeavor that evolved into the ICCF.[5] Next, I analyze how relationships among U.S.-based conservation NGOs, USAID, the U.S. Congress, and Malagasy state agencies have shaped the negotiations over the donor-funded NEAP and USAID's contributions to it. Finally, I concentrate on the effort to expand protected areas in Madagascar, an initiative funded by USAID. Collectively, these lenses reveal how the often well-intentioned action to attract high-level political attention to conservation by isolating the environment as a separate realm of governance has disconnected it from its embeddedness in human social, political, and economic relations and as a result constrained conservation approaches.[6]

I argue that the emergence of biodiversity conservation in foreign aid politics in the mid- to late 1980s resulted from the consolidation of state authority within global environmental institutions and national-level neoliberal reforms, which collectively reconfigured state, market, and civil society relations. These reforms reduced state management and enforcement capacity, liberalized economies, and encouraged foreign direct investment (FDI) while also bringing nonprofit and private actors into positions of authority in state policy-making processes. As the downsized Madagascar and U.S. agencies turned to these actors for assistance in policy processes, they became dependent on them to mobilize political support for their environmental programs.

This reliance impacted the Madagascar environmental program in four intertwined ways. First, the need to maintain strategic relationships with them—contrary to what neoliberal advocates might presuppose—reinforced upward accountability to decision makers in capital cities, undermining prior efforts to devolve resource management authority to rural communities. Second, the resulting public-private-nonprofit alliance of political champions successfully marshalled U.S. funding for conservation by isolating "the environment" geographically, as biodiversity "over there," threatened by Malagasy farmers conducting slash-and-burn agriculture, and best conserved by expanding protected area networks that, in the absence of sufficient state capacity, could be managed by private sector and nonprofit actors. Third, this framing, legitimized by appealing to global biodiversity targets, led to a narrowed environmental agenda that reflected the specific priorities of these advocates and offered politicians in both countries an avenue to become "green" without impeding the commercial resource extraction industries. Fourth, in focusing on the poor as instigators of deforestation and directing human, financial, and political resources toward expanding parks on paper rather than building state management and enforcement capacity to manage them, this alliance of actors collectively failed to create the necessary institutional capacity to stem expanding commercial resource extraction in the twenty-first century.

In this contribution to the interdisciplinary body of literature that critically examines "development," or the global agenda to improve the South, I explore why, after three decades and hundreds of millions of dollars in environmental aid, the Madagascar government still lacks the capacity to protect its environment, particularly given increasing pressure to exploit its natural resources. Illegal forest trade has been a known issue for decades, and the mining, oil, and gas industry is rapidly expanding and anticipated to attract increasing FDI in the coming decades. Mining and timber extraction already takes place in and around protected areas, and the rapid growth of protected areas on paper will do little to protect Madagascar's biodiversity in the absence of management capacity. It may even increase deforestation by diverse actors seeking claim to Madagascar's forests. Why, then, is so much attention focused on creating new protected areas?

Numerous scholars have attributed the strong biodiversity conservation program in Madagascar to the significant influence of U.S.-based organizations, from scientific institutions to conservation NGOs to aid agencies (e.g., Duffy 2006; Erdmann 2010; Horning 2008a; Kull 2014; Moreau 2008; Sarrasin 2007b). My goal is to document how this influence materializes not only through formal political negotiations and bureaucratic practice but also via informal, everyday interactions among people who span multiple geographic and institutional sites. In contrast to scholars who focus on USAID as a bounded organization, the politics of U.S. foreign aid more generally, or the impacts of USAID's programs (e.g., Berríos 2000; Butterfield 2004; Essex 2013; Hoben 1989; Lancaster 2007; Medley 2004; Mitchell 2002; Tendler 1975; Weissman 1995), I am interested in the transnational and interorganizational processes through which its programs are developed and implemented. I concentrate on the rise of a political alliance among representatives of USAID, the U.S. Congress, and U.S.-based transnational conservation NGOs.

In this regard, I do not provide a comprehensive overview of U.S. environmental foreign aid, the USAID Madagascar program, or Madagascar's environmental program. For one, I have underemphasized the agency and politics of the Madagascar government, other multilateral and bilateral donors, and development NGOs. In my endeavor to protect confidentiality I have also downplayed the critical roles played by specific individuals. My goal is neither to generalize about U.S. foreign aid nor to generate predictive theory or model political dynamics. Instead, by paying ethnographic attention to the individual relationships and informal events across multiple sites and scales that constitute U.S. environmental aid to Madagascar, I offer "close study of a particular part to generate broader claims and understandings" (Hart 2006b, 996) with the goal of informing theories of neoliberal conservation and environmental governance by reworking them to accommodate new findings (Burawoy 1998).

My analysis is situated within a rapidly growing scholarship embodied in the term "neoliberal conservation" that critiques claims to save the environment from the human impacts of capitalism while simultaneously espousing capitalism as the means to do so as well as analyzes the emergent ways in which nature is being commodified, privatized, and financialized (e.g., Brockington, Duffy, and Igoe 2008; Büscher et al. 2012; Büscher and Fletcher 2015; Corson, MacDonald, and Neimark 2013). Neoliberalism, with its ideological commitment to the free market, rose to prominence in mainstream economic policy in the 1980s, particularly under Margaret Thatcher in the United Kingdom and Ronald Reagan in the United States. In its various manifestations, it has tended to encompass privatization, marketization, deregulation, market-friendly regulation, use of market proxies in the remaining public sector, civil society provision of state services, and individual self-sufficiency (Castree 2008, 2010). Neoliberal reforms have not only "rolled back" the state and weakened its ability to regulate the private sector and to redistribute wealth but also "rolled out" new governing structures designed to facilitate the growth of enterprise (Peck and Tickell 2002). Yet, as numerous scholars have underscored, neoliberalism has been manifested in variegated and hybrid ways across the globe (e.g., Bakker 2010; Castree 2006; Larner 2003; Mansfield 2004; Peck 2004). Thus understanding "actually existing neoliberalism" (Brenner and Theodore 2002) requires examining the *process* of neoliberalization across multiple and "often connected, places and times" (Castree 2010, 1732).

In using the case of USAID's environmental funding in Madagascar to do just that, I resist the tendency in neoliberal conservation scholarship toward totalizing critiques (Castree 2014) as well as the emphasis in international development on evaluations of projects rather than of aid systems. I concentrate on how neoliberal reforms that reduced the state's and increase private and nonprofit organizations' involvement in public policy transformed not only who sat at the negotiating table but also which ideas became politically acceptable to voice in environmental politics. These shifting relations of governance led to what Goldman (2005) termed "green neoliberalism," or the convergence of liberal efforts to expand and intensify markets with movements for environmentally sustainable development in the global South, and, as I argue here, to the ascent of biodiversity conservation in foreign aid politics. By tracing how foreign aid for conservation has created new symbolic and material spaces for global capital expansion I underscore the relationship between what Hart terms "Big D Development," or "the post-Second World War project of intervention in the Third World emerging in a context of decolonization and the Cold War," and "little d development" as "the development of capitalism as a geographically uneven and contradictory set of historical processes" (2001, 650; see also 2006a). My goal is to understand the political work that neoliberal conservation does.

Drawing on the anthropologist David Mosse (2005), I contend that the prioritization of biodiversity conservation in both the Madagascar NEAP and USAID's environmental program in Madagascar has reflected efforts to mobilize and maintain *interorganizational* and *transnational* political support. Mosse asserts that the success of a development project is produced not through its impacts in the field but through its ability to stabilize critical social relations. Similarly, a number of political ecologists have argued that hegemonic environmental narratives constitute political maneuvers, which persist in the face of contradictory evidence precisely because they serve strategic purposes (Fairhead and Leach 1996; Fortmann 1995; Hoben 1995; Leach and Mearns 1996; Roe 1994). Unpacking hegemonic narratives and associated policies, then, requires attending to the strategic relationships they sustain and how they do this. In Madagascar, events such as conferences, and practices such as grant mechanisms have aligned various actors in the environmental program. Likewise, explanatory narratives have framed environmental problems in ways that engage these actors and that have justified their relationships. For example, the narrative attributing forest loss in Madagascar to rural peasants has been an effective political strategy to reaffirm public and private sector allegiances for centuries, and the claim that Madagascar was once completely forested has formed a "motivating *raison d'être*" for conservation interests to intervene in Madagascar resource politics (McConnell and Kull 2014).

The narratives and policies that persist both reflect and reinforce critical power relations among these actors. These relations are always historically contingent, dynamic, and conditioned by larger political-economic forces. In this regard, development paradigm shifts occur through complex, historically specific interactions and, as such, must be located within historical conjunctures and understood in relation to intellectual trends and shifts in global economic structures, political priorities, and institutional dynamics (Cooper and Packard 1997).

I begin with a brief introduction to forest politics in Madagascar and then trace the evolution of USAID's environmental program there, situating it within the broader global political economy transformation that has taken place over the past forty years; I also give brief snapshots of the chapters to come. I then discuss the importance of attending to the interorganizational relationships that comprise U.S. foreign aid, and I conclude with observations on the current state of Madagascar's environment.

Historical Struggles over Madagascar's Resources

Contemporary struggles over Madagascar's forests have been shaped by the country's combination of immense human poverty and great biological wealth. As the world's fourth largest island at 587 square kilometers and separated from the African continent by the 400-kilometer Mozambican channel, Madagascar is one of the poorest countries in the world. In 2012 the country's gross national income per capita was equivalent to US$473, and approximately 75 percent of its population lived below the poverty line. Poverty rates are highest in rural areas, where 70 percent of Malagasy residents live and where farmers depend on small-scale agriculture and forest products for their livelihoods (World Bank 2014). At the same time, it is one of the world's most biologically rich areas. Owing to its split from the Indian, Australian, and Antarctica subcontinents at least eighty million years ago, much of the flora and fauna of Madagascar has evolved in geographic isolation. It has extremely high species endemism: 80 percent of its species are found only in Madagascar, and as much as 90 percent of its endemic species live in forest ecosystems (Dewar and Richard 2012; Gorenflo et al. 2011; Wright and Rakotoarisoa 2003). While all regions of Madagascar contain important biodiversity, USAID has invested primarily in the eastern rainforests, where I focused my research.

As I explore in chapter 2, ideological and material struggles, for science, economic gain, and rural livelihoods, have shaped the use and management of Madagascar's eastern rainforests for centuries. These struggles have created the political, economic, and discursive conditions under which modern policy makers have approached forest conservation in a number of ways. First, successive governments— from the precolonial Merina Empire, which ruled Madagascar from the sixteenth century to the nineteenth, to the French colonial government to the postcolonial state—have all pursued policies designed to maintain control of Madagascar's forests that created incentives for shifting cultivators and large-scale commercial operators alike to deforest them. They have blamed shifting cultivation, or *tavy*, as it is called in Madagascar's eastern forests, for deforestation and prohibited it, even as they have promoted forest clearance for domestic industrialization and trade and granted foreign traders land rights for mining, timber, and cash crop plantations.[7] They have also claimed forestland but recognized individual rights to cultivated land. In the face of enclosures of peasants' lands, tavy has been an effective means to turn land owned by the state into cultivated land to which the government has recognized individual rights (Bertrand, Ribot, and Montagne 2004; Harper 2002; Jarosz 1993; Keck, Sharma, and Gershon 1994; Keller 2008; Kull 2004; Sodikoff 2005). Second, lack of human and financial resources has meant that the state was unable to contain deforestation by commercial exploiters, an issue that continues to this day. By reaffirming state ownership of forestland and restricting villagers' forest-based activities, while simultaneously condoning large-scale commercial exploitation, contemporary conservation efforts are reinforcing historical incentives to claim land by clearing it. Importantly, however, current conservation efforts are being implemented under conditions of changing power dynamics among public, private, and nonprofit entities. In the wake of neoliberalism, forest policy in Madagascar is no longer just a state project.

Scientific attempts to classify Madagascar's species were recorded as early as the mid-seventeenth century, and formal scientific explorations from Europe took place in both the eighteenth and nineteenth centuries as well as during the French colonial era (Anderson 2013; Andriamialisoa and Langrand 2003; Feeley-Harnik 2001). These early expeditions and classification efforts laid the foundation for subsequent research endeavors and conservation approaches. The French colonial government established a legacy of separate "conservation areas" and "production zones" that sought to isolate priority conservation zones from production pressures. Malagasy and French research institutes supported scientific expeditions both before and after independence in 1960. However, from what is referred to as the "second revolution" in 1972, which sought to further remove French influence in political and economic affairs, until the 1980s the government discouraged foreigners from conducting research in Madagascar (Andriamialisoa and Langrand 2003; Fenn 2003; Jolly and Sussman 2007). In light of these restrictions and in response to growing concerns about forest loss, a core group of scientists, many of whom were American or associated with American institutions, organized a series of meetings with Malagasy policy makers to develop institutional protocols for continuing research and to urge the conservation of Madagascar's flora and fauna. The relationships that developed among the actors who attended these meetings as well as the policies and conservation priorities they enacted continue to influence Madagascar conservation politics. The critical historical conjuncture of the mid-1980s, with its emphasis on biodiversity, sustainable development, and neoliberalism, provided the political-economic context for these dedicated actors to transform a scientifically grounded movement into an institutionalized donor-funded program.

Shifting Relations of Environmental Governance Under Neoliberalism

Guided by Keynesian principles, much environmentalism in the 1970s hinged on public confidence in the state to regulate private activities and to protect human welfare, in natural resource limits to growth, and in the need to redress the negative environmental effects of capitalism (McCarthy and Prudham 2004; Meadows et al. 1972; UN 1972). In international development, foreign aid agencies had begun shifting their investments away from large capital projects, and the announcement by the president of the World Bank Robert McNamara in 1973 of his decision to tackle rural poverty launched the "basic needs" development era. That same year the U.S. Congress passed the New Directions legislation, which required USAID to prioritize agriculture, health, education, and population planning in developing countries (U.S. Congress 1973). While the legislation did not specifically focus on the environment, it established a foundation for expanding USAID's programs to environmental issues. In the mid- to late 1970s a group of environmental advocacy organizations convinced Congress to amend the Foreign Assistance Act (FAA) to require USAID to conduct environmental impact assessments of its projects and to fund environmentally focused programs. The agency's first environmental projects emphasized state intervention to manage natural resource supplies for the poor (U.S. Congress 1977, 1978, 1981a; see also USAID 1988c).

Nevertheless, the roots of the neoliberal revolution were already forming in the 1970s, as evidenced by the Chicago economists' restructuring of the Chilean economy according to neoliberal economic ideals; the delinking of the U.S. dollar from the gold standard and abandonment of fixed exchange rates; and the resulting recycling of petrodollars through New York investment banks to third world governments (Gore 2000; Gowan 1999; Hart 2006a; Harvey 2005). While Paul Volcker began the U.S. government's initial move toward neoliberal monetary policy during the Carter administration, Reagan further deregulated various industries, reduced corporate taxes dramatically, and promoted the reduction of government and the expansion of the private sector.

Neoliberal champions blamed foreign aid failures on Keynesian state-coordinated development (Auer 1998; Berríos 2000; Essex 2013).

Provoked by the lending of petrodollars and the subsequent hike in interest rates in the 1970s, a debt crisis had hit the global South by the 1980s. Many countries, including Madagascar, acquiesced to the International Financial Institutions' (IFI) structural adjustment reforms, and bilateral and multilateral development agencies endorsed the Washington Consensus approach, which prioritized fiscal austerity, trade liberalization, privatization, and deregulation (Gore 2000). Facing an economic crisis, the Madagascar government began abandoning its nationalist and socialist agenda, pursuing instead stabilization and liberalization policies in the early 1980s. It devalued its currency, reduced government expenditures, lowered trade barriers, privatized state monopolies, and encouraged foreign investment (Barrett 1994; Kull 1996; Mukonoweshuro 1994). By the mid-1980s the Madagascar government's agreement to structural adjustment had prompted a rapid influx of Western donor assistance. Concomitantly, environmentalists were pushing development donors to fund environmental programs under the auspices of the sustainable development discourse, and species conservation advocates had initiated a campaign to protect species and habitat diversity around the world, encompassed in the newly coined term "biological diversity" (Takacs 1996).

The rise of neoliberal policies in Madagascar was accompanied by new approaches to the study of the country's flora and fauna and new requirements to protect them. In 1980, the year Madagascar signed its first stabilization agreement with the International Monetary Fund (IMF), the World Conservation Strategy endorsed economic development as a means of achieving conservation rather than seeing it as a threat to conservation. Even as sustainable development directly contradicted the 1970s stress on limits to growth, the idea that economic growth and environmental conservation could be compatible spread rapidly because it offered a political avenue for environmentalists to engage in international development politics (Adams 2008; Hajer 1995; Redclift 1992), and many environmental organizations had already endorsed this idea by the time the Brundtland Report introduced the oft-quoted definition of sustainable development as "development which meets the needs of the present without compromising the ability of future generations to meet their own needs" (UN 1987). Environmental NGOs used the concept of sustainable development to urge the U.S. Congress to make additional environmental amendments to the FAA bill, and by the early 1980s USAID's environmental program was expanding rapidly, with congressionally mandated programs specifically on tropical forest conservation, endangered species, and biodiversity.

The discipline of conservation biology was founded in the 1980s, encapsulated in the creation of the Society for Conservation Biology and the coining of the term "biodiversity" at the National Forum on BioDiversity in 1986. That same year the U.S. Congress began setting aside funds for biodiversity conservation in the appropriations bill, and it directed USAID to channel these funds to NGOs. This shift was critical: while early congressional language pushing environmental programs had *authorized* funds, at this turning point the Congress began actually *appropriating* money—or requiring that USAID spend money—for biodiversity conservation.[8]

Citing the government's continued adherence to IFI reforms, USAID started expanding its programs in Madagascar in the early 1980s, focusing first on agricultural productivity, infrastructure, and economic growth and then, in 1987, as it responded to congressional biodiversity mandates, on Integrated Conservation and Development Projects (ICDPs). In contrast to the Yellowstone Park Model, which had embraced the idea that nature was best preserved by removing it from human influence (Neumann 1998), ICDPs incentivized conservation by giving development aid to people living around protected areas. Together with CBNRM, which provided economic benefits and devolved management responsibility to communities, ICDPs spread across the developing world in the late eighties and nineties, funded to a great extent by USAID (Adams and Hulme 2001; Alpert 1996).

As USAID partnered conservation NGOs with development organizations to implement ICDPs, it brought the conservation groups into foreign aid politics:

> The big donors knew little about sustainable development, but they wanted to look green. Conservation organiza-

tions, impeccably green, but inexperienced in managing international development projects, wanted to expand
their programs and their international reach. Through
multimillion-dollar USAID projects, conservation groups
could demonstrate impressive rates of financial growth to
their boards of directors. The self-interest of the two sets of
organizations converged in [ICDPs]. For USAID and its
counterparts, ICDPs provided a green shield under which
they could continue what they had been doing for decades,
sponsoring rural development programs in developing
countries. For the conservation organizations, ICDPs offered a path to rapid growth. (Terbough 1999, 164–65)

In Madagascar USAID's initial proposals for an expanded environmental program invoked conservation NGO reports about the
plight of Madagascar's flora and fauna as justification, and it funded
conservation and development NGOs to implement its initial
program.

Launching Madagascar's Environmental Program

Madagascar's overarching environmental program emerged out of
the combination of neoliberal development planning and the rise
of donor funding for the environment. In 1987, in response to
NGO pressure, the World Bank president Barber Conable announced his intention to place more attention on environmental
issues (Goldman 2005; Rich 1994; Wade 1997), and it began looking for countries in which to launch more comprehensive environmental programs. Because the Madagascar government had begun
structural adjustment reforms, giving the World Bank political leverage, and because scientists and policy makers had already built
strong collaborations around conservation, Madagascar offered an
ideal place. The resulting donor- and government-funded fifteen-
year, three-phase NEAP in Madagascar brought together donors
and government around the aim of integrating environmental policy into the island's overall development plans; the World Bank
later referred to it as "the most ambitious and comprehensive environmental program in Africa" (World Bank 2007, ix). Even as the

NEAP proposed a broad agenda on paper, the vast majority of funds for it came from foreign donors, and as a result its priorities gradually shifted to reflect those of the donors. While the Madagascar government sought to push funds toward rural development, U.S.-based donors and NGOs directed sizable funds to biodiversity conservation. USAID financed the first NEAP donor-coordinating group, funded the development of environmental institutions, and eventually became the second largest donor (after the World Bank) to the NEAP over its lifespan (World Bank 2007).

The conceptualization of Madagascar's NEAP at this historical moment shaped the realm of possibilities for its subsequent agenda. An emphasis on biological diversity, a reduced role for the state, and a strong role for NGOs informed the politically viable narratives used to frame Madagascar environmental challenges, the strategies that could be invoked to redress them, and the actors granted the authority to manage its resources. First, the World Bank recruited scientists, NGOs, and consultants to help develop and implement the program's priorities, thereby setting a precedent for involving nonstate actors in state environmental deliberations (Duffy 2006; Falloux and Talbot 1993; Kull 1996; Sarrasin 2007b). Second, the IFI push to downsize the Madagascar state, combined with donor skepticism about corruption in the Madagascar forest service, led to the creation of new independent organizations through which donors could channel environmental funds. However, as donors circumvented established government agencies, they reinforced the state's dependence on external actors for program design, management, enforcement, and funds. Finally, the NEAP embraced a reduced state and liberalization of the economy as means to achieve its goals (République Démocratique de Madagascar 1990). Resulting government policies tried to balance liberalized trade in natural resources with conservation by focusing on establishing protected areas in order to safeguard biodiversity from the actions of rural peasants, while simultaneously reducing taxes and strengthening investor rights in order to attract FDI in natural resource extraction. In policies ranging from the environmental charter that accompanied the NEAP to Ravalomanana's later policies, Madagascar Naturellement (Naturally!) and Madagascar Action Plan (see chapter 5), conservation became a politically palatable way to ensure that the environmental agenda did not impede economic growth.

The agency's increasing partnership with NGOs reflected a shift toward a softer form of neoliberalism that had appeared by end of the 1980s. At the World Bank this meant embracing environment, participation, and social capital as well as what Joseph Stiglitz termed the "*post*-Washington Consensus," which called for an emphasis on improving living standards, promoting sustainable development, correcting market failures, limiting regulation, and controlling short-term international capital flows (Gore 2000; Hart 2001). Foremost in this modified version of neoliberalism were idealized visions of NGOs as embodiments of civil society, alternatives to failing states, more efficient and cost effective than governments, and a counterweight to the influence of the private sector. Embracing this vision, multilateral and bilateral donors turned increasingly to NGOs to implement foreign assistance. Bilateral aid directed through NGOs rose from 0.7 percent in 1975 to 5 percent in 1993–94 (US$2.3 billion in absolute terms) (Edwards and Hulme 1996). Riding this wave, environmental NGOs across the globe became conduits for new state-sponsored sustainable development programs (Edwards and Hulme 1996; Smillie 1997; Zaidi 1999).

In Madagascar, as the agency scaled up funding for ICDPs and environmental infrastructure during the first phase of the NEAP, it began channeling millions of dollars through U.S.-based conservation NGOs operating in Madagascar. This funding and associated political support facilitated the NGOs' transformation from environmental researchers and advocates to program managers and then to influential political players in Madagascar conservation politics. Concurrently in Washington, D.C., as the environmental advocacy organizations that had originally lobbied for USAID's environmental programs turned to focus on the World Bank, the conservation NGOs funded by USAID became the primary political advocates for USAID's environmental programs on Capitol Hill. Both the U.S. Congress and the Clinton administration saw NGOs as countering private operators and urged USAID to channel biodiversity funds through NGOs. This idealized vision of NGOs continued into the twenty-first century even as they began developing corporate partnerships.

By the early 1990s global environmental discourses associated with the rise of international environmental institutions had recast environmental degradation as the result not of growth but of policy failures that could be corrected through market solutions (McAfee 1999). In 1992 the United Nations Conference on Environment and Development in Rio de Janeiro (the Earth Summit) embraced sustainable development as its core theme, creating a post-summit Commission on Sustainable Development (CSD) in addition to conventions on biological diversity, climate change, and desertification. Even as these conventions consolidated state environmental authority in multilateral agreements, they institutionalized long-term participatory mechanisms for involving nonprofit and private actors in future negotiations, such as the CSD Major Groups process. They subsequently became focal points for actions by governments, the private sectors, NGOs, and a new generation of transnational environmental movements brought together by technology and globalization (Corson et al. 2015; Keck and Sikkink 1998; Tarrow 2005). They also became arenas for articulating the environment as a "global good" (Taylor and Buttel 1992).

Elected immediately after the Earth Summit, the Clinton administration advocated the use of foreign aid to protect the global commons, emphasizing global environmental problems such as climate change and biodiversity conservation (Lancaster 2007; USAID 1996, 1997a). Nonetheless, both the Clinton administration and the Republican Congress implemented neoliberal reforms that dramatically reduced USAID's financial and personnel resources and turned over much of USAID project management to contractors and grantees. These budget pressures translated into the closing of missions around the world, reinforcing the need for missions to maintain strategic alliances with Washington-based USAID, congressional, and NGO staff in order to protect their funding.

By this time conservation NGOs had grown in size and influence, fueled by increasing USAID funds and by their expanding collaborations with multinational corporations, the financial sector, and the entertainment industry (Birchard 2005; Hiar 2013, 2014; MacDonald 2008; Ottaway and Stephens 2003). As they grew they became critical advocates for USAID. In 2003 WWF-U.S., WCS, CI, and TNC created an International Conservation Partnership (ICP), which aimed to build congressional support for conservation and which evolved into the ICCF. The group attracted corporate partners and a diverse set of congressional members by focusing primarily on foreign biodiversity and invoking an anti–big government message even as it pushed for more public expenditures on conservation. In doing so, it created a way for otherwise antienvironmental politicians and corporate leaders to be green without addressing difficult domestic environmental issues or the degradation caused by their own companies. With this broad political and corporate backing, the ICCF was influential on Capitol Hill. When the George W. Bush administration reduced USAID's environmental investments, the ICCF and associated conservation NGOs advocated not only to preserve biodiversity funds but also to protect the USAID Madagascar mission from funding cuts. As overall environmental funding declined, the biodiversity funds became one of the few ways USAID could continue environmental programs.

The Conservation Enterprise

These changing relations of power and authority among public, private, and nonprofit organizations were intertwined with transformations in the practice of conservation, specifically the rise of market-based approaches, science-based priority setting, and conservation planning at a landscape scale. By the beginning of the twenty-first century, participatory conservation had given way to market-driven programs, such as payment for ecosystem services and biodiversity offsets. The introduction of the internet and the financialization of markets had led to the production and circulation of virtual commodities, in which buyers could purchase carbon offsets and wildlife derivatives (Arsel and Büscher 2012; Igoe 2010; Sullivan 2013). These market-based programs, focused on project-level interventions, were designed to offset extractive processes, enabling them to continue rather than challenging them. For example, in Madagascar USAID financed feasibility studies for a project to fund conservation through carbon credit sales in the

Makira protected area (Brimont and Bidaud 2014; Ferguson 2009; Méral et al. 2009); QIT Madagascar Minerals launched a program to offset its environmental impacts by creating new protected areas (Kraemer 2012; Seagle 2012; Waeber 2012); and WWF, CI, and the Madagascar government established the Madagascar Biodiversity Fund, an investment fund that by 2014 had raised over $US50 million (Madagascar Biodiversity Fund 2015; Méral et al. 2009; Méral 2012).

At the same time, drawing on the science of island biogeography and relying on geographic information systems (GIS), conservation advocates had begun arguing that conservation was best accomplished not via isolated parks or community projects but via ecoregional and transboundary efforts (e.g., Attwell and Cotterill 2000; da Fonseca et al. 2005; Margules and Pressey 2000). In extreme cases, some advocates called for a return to exclusionary conservation (Oates 1999; Terbough 1999), a move that has been termed "back to barriers" (Hutton, Adams, and Murombedzi 2005; see also Wilshusen et al. 2002). The idea of ecoregional planning appealed both to strict conservationists who wanted to use it to expand protected areas and to those who saw it as a means of addressing nonlocal drivers of environmental degradation. It empowered organizations with strong GIS skills that could illustrate priorities at a regional scale. USAID adopted the concept of ecoregional planning during the second phase of the NEAP, using it to promote both integrated land use planning and protected area networks.

As international environmental institutions like the Convention on Biological Diversity (CBD) and IUCN adopted conservation planning at a landscape scale, they endorsed numerical targets for protected area coverage. In 1992 the Fourth WPC recommended that protected areas cover at least 10 percent of every biome by 2000 (Langhammer et al. 2007), and the Millennium Development Goals and the CBD subsequently adopted similar versions of this aim (CBD Secretariat 2007; UN Statistics Division 2008). By 2005, 12 percent of the earth's terrestrial surface was preserved (Adams and Hutton 2007; Chape et al. 2005), with the coverage of specific biomes varying from 4 to 25 percent (Jenkins and Joppa 2009; see also Brooks et al. 2004; Hoekstra et al. 2005). In 2010, after much lobbying by CI and other organizations, the Conference of the

Parties to the CBD agreed to increase its protected area target to 17 percent of terrestrial areas and 10 percent of coastal and marine areas (Campbell, Hagerman, and Gray 2014; Corson 2011; World Bank 2011a).

These highly visible and easily communicated metrics have reflected and reinforced global claims to the authority to define what "conservation is, how it will be accomplished, and who is responsible for it" (Campbell, Hagerman, and Gray 2014, 60). They have undermined the decision-making autonomy of local users and legitimized new expropriations of land and resources under the guise of protecting "global goods." The associated networks of protected areas—as they have created and mapped new boundaries, allocated and enforced new rights, and designed acceptable resource uses—have entailed processes of territorialization (Peluso and Vandergeest 2001; Vandergeest and Peluso 1995; Vandergeest and Peluso 2006a, b). Ultimately, by alienating some actors of their land and resource rights while securing rights for others, they have embodied accumulation by dispossession (Corson and MacDonald 2012; Harvey 2003; Kelly 2011), or "green grabs," in which "green credentials" are used to justify the alienation of already marginalized groups of people from their rights to and control over resources (Fairhead, Leach, and Scoones, 2012, 237; see also Borras, McMichael, and Scoones 2010; Borras et al. 2011; Peluso and Lund 2011).

In Madagascar, alliances of public, private, and nonprofit actors secured the authority to create new boundaries, resources rights, and acceptable uses as they endeavored to meet Ravalomanana's target. The resulting protected areas created new opportunities for capital accumulation by privatizing parks, promoting new commodities such as Reducing Emissions from Deforestation and Forest Degradation in Developing Countries (REDD+) and biodiversity offsets, accommodating mining interests, and enabling donors, government, NGOs, and media companies to attract new investment by showcasing Madagascar's park expansion as a conservation success (Corson 2011; Corson and MacDonald 2012). Collectively these actions have delegitimized the power and authority of rural peasants to control the source of their livelihoods and to garner wealth from forest resources.

Enclosures in the name of conservation are not new: scholars have documented displacements of local and indigenous peoples under the guise of biodiversity conservation across the globe (e.g., Brockington, Igoe, and Schmidt-Soltau 2006; Brockington, Duffy, and Igoe 2008; Chapin 2004; Neumann 1998; Peluso 1993). However, what is new are the dispersion of authority to decide acceptable resource uses within targeted areas across local, national, and international scales and the transnational networks of state and nonstate actors (Corson 2011; Igoe and Brockington 2007; Sikor and Lund 2009). Networks of conservationists, philanthropists, celebrities, private companies, and others have raised enormous sums for conservation through galas and exotic trips (Holmes 2011, 2012), creating what I have termed a conservation enterprise, in which funds are shifted among public, private, and nonprofit entities without ever being used on the ground (Corson 2010). Through short trips and galas, congressional members and corporate executives have developed a sense of familiarity with environments over there, and by focusing their green efforts on biodiversity conservation they have avoided facing the potential negative impacts of their own organizations' actions, whether U.S. trade policy or unsustainable resource extraction practices. In Madagascar, by means of active participation in the working groups created to implement SAPM, foreign aid donors, consultants, scientists, and national and transnational conservation NGOs have undertaken functions that would otherwise have been state responsibilities, namely, writing laws, developing maps, and establishing boundaries that were subsequently issued officially by the Madagascar government. As these assemblages of public, private, and nonprofit actors have negotiated territorial boundaries and associated resource rights, they have governed these spaces according to their own needs, not those of local actors (Igoe and Brockington 2007).

Ethnography of Relations Across Time and Space

Understanding the diffuse forms of power embodied in these transnational networks and their effects requires moving beyond studies of single organizations and bounded sites. It also necessitates paying ethnographic attention to both formal and informal domains in which dynamic groups of actors come together. In my endeavor to take up these challenges, I join the growing number of ethnographers who have extended the application of ethnography beyond its original use in social and cultural anthropology in order to "study up" (Nader 1972) to capture the micropolitics of the purportedly "all-powerful development institutions" (Watts 2001, 286; e.g., Crewe and Harrison 1998; Goldman 2005; Lewis et al. 2003; Lewis and Mosse 2006; Mosse 2005) as well as those who critically assess the increasing influence of NGOs in foreign aid policy (e.g., Fisher 1997; Lewis and Opoku-Mensah 2006; Lister 2003; Zaidi 1999).

By directing attention toward the constitutive and transnational processes through which foreign aid is continually negotiated, I understand policy making not as autonomous from practice (van den Berg and van Ufford 2005) but as everyday practice. In contrast to actor network analyses that combine Foucauldian and actor-network theory (e.g., Fairhead and Leach 2003; Lewis and Mosse 2006; Mosse and Lewis 2005), I focus on the messy, dynamic process of compromise that characterizes policy making, and I emphasize how and why particular processes come together at specific historical conjunctures (see also Büscher 2013; Goldman 2005; Hart 2002; Li 2007). This approach underscores that power is not inherent in structures but relational and dynamic—continually formed, maintained, and contested through interaction—and productive of structures as well as reproduced through them. It reveals foreign aid as a continually contested process rather than a monolithic, all-powerful, Western-imposed development machine, epitomized in James Ferguson's book *The Anti-Politics Machine*. Ferguson argued that the World Bank aid to Lesotho, through its apolitical rendering of poverty as a technical problem, became "a machine for reinforcing and expanding the exercise of bureaucratic state power" (Ferguson 1994, 180). While the postdevelopment literature brought scholarly attention to the ways in which the dominant development discourse produces permissible modes of being and thinking while disqualifying others (e.g., Escobar 1995; Ferguson 1994; Sachs 1992), I build on an alternative body of scholarship that rejects both structural and discursive determinism and focuses instead on the micropolitics through which global development discourses are challenged, refracted, and reworked (e.g., Gupta 1997; Harrison 2003; Li 2007; Moore 1999). Emphasizing how USAID and the Madagascar environmental program are dynamic, constantly being reformulated by devoted individuals, many of whom are trying to challenge the system even as others thwart their efforts, I argue that moments such as brainstorming sessions in a hotel room can be as influential as formal policies and negotiations.

In contrast to many institutional ethnographers, I also examine the development bureaucracy as an unbounded entity, seeking to move beyond dichotomies that posit spaces as inside or outside organizations (Billo and Mountz 2015) and relying instead on the use of multisited ethnography to trace the movement of development concepts, programs, and politics across transnational networks (see also Bebbington and Kothari 2006; Bebbington et al. 2007; Goldman 2005; Moore 2001; Perreault 2003; Seidman 2001; Thayer 2001). The core challenge in such a method is determining how to bound the ethnographic field in time and space.

Rather than defining my ethnographic field of study by following people or things (Gusterson 1997; Marcus 1995, 1998), concentrating on relationships among sites (Hannerz 2003) or constellations of actors (Wedel et al. 2005), or tracing "distended networks" (Peck and Theodore 2012), I turn to Massey's understanding of place, not as static or bounded but as a node of connection in socially produced space. Applying it to policy-making sites, I examine how various state and nonstate actors come together at geographically and temporally specific moments. For Massey (1999), "Places may be imagined as particular articulations of these social relations, including local relations 'within' the place and those many connections which stretch way beyond it. And all of these [are] embedded in complex, layered histories" (41). Thus events such as WPC, moments such as presidential announcements, and institutions like the ICCF can all be understood as bundles of social relations and power dynamics that are formed through historically and geographically sedimented practices and processes.

In order to theorize which nodes become pivotal—or how "power relations and historical sediments formatively shape contingent constellations that become materially and discursively consequential" (Moore 2005, 25)—I use Hall's (1985, 1996, 2002) concept of articulation as both the joining together of diverse elements and expression of meaning through language. Drawing on Gramsci's understanding of hegemony as a terrain of struggle, or the "compromises through which relations of domination and subordination are lived" (Gramsci 2010 [1971]), Hall (1985) underscores that connections are not given but require particular conditions to exist and that rearticulations are constantly being forged. Accordingly, I examine USAID conservation policy as a negotiated terrain of contingent, temporary, and fragile articulations that are made in historically specific contexts, emerge through practice, and are subject to continual reworking. Over time and space, these articulations collectively configure a transnational field of foreign aid governance (see also Corson, MacDonald, and Campbell 2014; MacDonald and Corson 2014), a field that is constituted across multiple sites and organizations and through multiple corridors of power.

Research Methods

In ethnographies of political institutions, the immersion needed to gain a situated understanding of the internal dynamics of a studied culture is often hindered by difficulty in accessing centers of power or the need to protect confidentiality in order to maintain that access. Institutional ethnographers, broadly defined, have necessarily combined key informant interviews, focus groups, document analysis, oral histories, and financial flow analysis with participant observation in order to produce the equivalent of "thick description" (Geertz 1973, 1988) in circumstances in which opportunities "classic participant observation" are restricted (e.g., Bebbington and Kothari 2006; Bebbington et al. 2007; Lewis et al. 2003; Markowitz 2001). In my case, long-term immersion in some of the studied organizations fundamentally shaped my research design, process, and final analysis. As a former employee of USAID and the U.S. Congress as well as of the White House, the World Bank, and the British Department for International Development, I spent a decade working at various levels and on different aspects of foreign aid and environmental policy prior to undertaking this research. These experiences not only afforded me personal

contacts but also gave me a familiarity with many of the bureaucratic practices and political dynamics I was studying.

I first visited Madagascar in 2000 as a USAID policy analyst,
after being intrigued by the USAID Madagascar environmental officer's lunchtime seminar on the challenges of simultaneously addressing poverty and environment in Madagascar. In my early
years of graduate school I collaborated with Malagasy and
American consultants on an evaluation of the Ampatsy CBNRM
project in the Alatsinainy Ialamarina rural commune, and I returned to Madagascar in 2005 and 2006 to conduct my dissertation
research.

However, because I was not engaged professionally in many of
the negotiations I studied for this research, I was not privy to the
hidden transcripts that shaped them. Therefore, like Tsing (2005),
in compiling this analysis, I put together "ethnographic frag-
ments." I used participant observation of public meetings, U.S. and
Madagascar government policy documents, personal and colonial
archives, and interviews of key informants, and I employed interviews to reveal the "back stories" and "hidden transcripts" behind
documents.

Hundreds of documents—including Madagascar government
legislation, USAID policy and program documents, USAID
congressional presentations, congressional appropriations and
authorization bills, hearing records, SAPM committee policy statements, conference reports, project evaluations, news releases, and
NGO lobbying material—afforded me historical and background
information.[9] I relied to a great extent on the physical and virtual
collections of the U.S. Library of Congress. I also conducted
extensive archival research in Madagascar at the Bibliothèque
Grandidier (Grandidier Library), the Madagascar National
Archives, and the Académie Malgache (Malagasy Academy) as well
as at the libraries of Landscape Development Interventions (a
USAID project) in Fianarantsoa and those of the USAID mission,
the World Bank, the Centre de Coopération Internationale en
Recherche Agronomique pour le Développement (CIRAD)
(Center for International Cooperation in Agronomic Research for
Development), and the Direction Générale des Eaux et Forêts
(DGEF) (General Directorate of Water and Forests), all located in

Antananarivo. Finally, I was privileged to use the letters, draft policy documents, diaries, and other treasures held in the extensive personal archives of Alison Jolly.[10] Her detailed notes and unique copies of historical documents offered invaluable insights into the personal relationships that comprised the scientific effort to bring international attention to Madagascar's biodiversity and the political negotiations behind the NEAP.

In Madagascar I gained basic background information from meetings among NGOs, donors, government officials, villagers, scientists, and others to which I was invited. These related to such issues as biodiversity prioritization, the integration of rural development and environment, and community consultation and forest governance. In villages I also attended focus groups with village leaders and representatives to talk about community conservation and the parks initiative.

Finally, I gained critical insights into individuals' motivations and goals as well as the historical context for my observations via 214 interviews that I conducted between 2003 and 2010 in Washington, Lewes and Cambridge, England, Antananarivo, the regional cities of Toamasina and Fianarantsoa, and selected villages within the Ankeniheny-Zahamena and Fandriana-Vondrozo biological corridors in Madagascar's eastern rainforest. I interviewed a range of actors, including current and former representatives of USAID; the U.S. Congress; national, regional, and local branches of the Madagascar government; other bilateral and multilateral foreign aid donors; transnational conservation and development NGOs; Malagasy NGOs; private sector companies; consultant groups; scientific organizations, including universities; and lobbying organizations based in Washington, D.C. I selected interviewees through a combination of targeted and snowball sampling, in which I asked all of them to recommend other potential candidates. From the resulting ever-increasing list I tried to interview people with a wide range of perspectives, organizations, and bureaucratic levels (from agency leaders to program staff) as well as those who could offer extensive historical information or who personally directed lobbying efforts, wrote particular pieces of legislation, or oversaw USAID's environment program. Many of the interviewees were current or former senior officials, and, as is typical, many spoke from experiences in more than one relevant position or organization. All Malagasy and French interviews and most English interviews were digitally recorded and then translated. Most interviews conducted were in person, but a few were by telephone. Interviews with two mayors and members of three Communautés de Base (COBAs) (Natural resource management community associations) and four villages in Fianarantsoa and Toamasina Provinces were conducted with the assistance of translators and facilitators.

In this book, I use the voices of interviewees to bring alive the negotiations that constitute the bureaucracy. However, because of the sensitive nature of some of the interviews, all information is reported anonymously in that sources are identified only by general position. My purpose was not to critique particular individuals' actions but to reveal the complex processes through which policy actions transpire.

Implications for Conservation and Livelihoods

In the following chapters I use multiple entry points to illustrate the historical, political, and intellectual corridors through which struggles over Madagascar's resources are intertwined with U.S. domestic politics. Here, ethnographic insights across sites and scales reveal how the everyday decisions of donors, NGOs, consultants, government officials, and others collectively created, contested, reshaped, and reinforced particular relations of environmental governance. By attending to the dialectical relationship between individual agency and structural power I hope to illustrate how the rise of neoliberalism brought particular ideas about conservation and relations of conservation governance to the fore and also how diverse groups of actors both contested and reproduced the structures that constrained them. As private and nonprofit actors became active partners in state policy processes in the wake of modified neoliberalism, they pursued environmental policies that could be implemented in the context of state reduction, liberalization of the economy, and efforts to attract FDI.

Biodiversity conservation offered a political pathway through the contradictions of sustainable development and green neoliberalism. The expansion of protected areas became a means by which to retain a commitment to foreign conservationists, while also promoting the extractive industries that were critical to the economy. Although the framing of Madagascar's environmental issues as a peasant problem to be solved by expanding protected areas was an effective political strategy to engage critical political allies, it directed human, financial, and political resources away from shaping national economic policy discussions and building the state and community management capacity needed to sustainably manage Madagascar's resources. At best, the expanded park network will contain parks that exist only on paper, and at worst it will catalyze increased deforestation as various actors, from rural peasants to mining prospectors, compete to stake claims to land and resources. By blaming rural peasants for deforestation and restricting their commercial activities in order to establish protected areas, while concurrently facilitating large-scale commercial resource exploitation, conservation approaches are reproducing historical conflicts that create incentives to appropriate land before someone else does. The stakes for Madagascar's biodiversity and people are high, especially given the expanding mining, oil, gas, and forest industries. Even as there have been investments in forest sector reforms and mining governance, policies to reduce regulation, promote exports, and encourage FDI are all designed to increase natural resource exploitation and allow the majority of wealth from these resources to leave the country. The key challenge that faced policy makers was not to create more protected areas but to build a domestic institutional infrastructure that could effectively manage those that the country already had.

By writing a critique of neoliberal conservation grounded in a detailed analysis of U.S. foreign aid to Madagascar, I hope to challenge readers to understand conservation as a political intervention that orders relations of power, in effect recasting winners and losers in struggles to control not just land and resources but also the ways in which certain actors can garner survival and wealth from them. Conservation is embedded in and productive of transnational, interorganizational relationships that are continually reshaped and reinforced in everyday hallway encounters as well as via official policy. Thus gathering more information, redesigning projects, consulting with communities, or passing new laws cannot rectify ineffective projects. The political relationships that they sustain and that have sustained them must also be changed. Transforming conservation in Madagascar will require not only modifying projects, programs, and funding streams but also redressing the transnational power relationships across scales, relationships through which both the current environmental crisis of the island and efforts to redress it have been created. Ultimately, such a reworking of power relations will require redistributing not just rights and access to forest resources but also decision-making authority in conservation governance.

CHAPTER TWO

The History of Forest Politics in Madagascar

Now that all the forests have become state property, they belong to he who makes them go up most quickly in flames. . . . Therefore, let us destroy the forest. A property that is not ours is thus replaced by one that does indeed belong to us.

—ABEL PARROT, 1925

If we compare this forest surface to the effective staff responsible for protecting and valorizing the forest, we see clearly the impossibility that the forest service could effectively accomplish its mission.

—LOUIS LAVAUDEN, 1934

Structural inequalities and the criminalization of *tavy* in protected forests lead rice farmers to think that *vazaha*

(Europeans, outsiders) and elite state officials use
"conservation" as a means to appropriate forests.

—GENESE SODIKOFF, 2005

The conservationist program is an assault on one of the most
fundamental values held by people in rural Madagascar: that
is, the value of the growth of life through kinship and
through one's roots in the land. In many ways, therefore,
conservationists ask the Malagasy to give up the very ethos
on which their lives are based.

—EVA KELLER, 2008

THE WORDS OF THE four authors quoted in the epigraphs above
offer complementary views on the complex, intertwined historical
processes that have generated Madagascar's contemporary environmental crisis and shaped proposed solutions to it.[1] In this chapter I
trace the intersection of land tenure, forest policies, and conservation efforts from precolonial to postcolonial times, showing how
they created the conditions under which specific ideas and narratives about forests as well as political and social relations around
their use and management emerged. I argue that it is critical to
understand this history, as contemporary conservation approaches
are reproducing prior conflicts over Madagascar's forests. By blaming shifting cultivators for deforestation and expropriating their
resource rights while also facilitating domestic industrial and
export-oriented exploitation, they are reinforcing incentives to
claim forested land by clearing it.

For hundreds of years small farmers, the state, commercial
exploiters, and conservationists have claimed and counterclaimed
control over Madagascar's eastern rainforests for shifting cultivation, international trade, and species protection. In this conflict
foreigners have introduced ideas about conservation, forest management, and agriculture that have intersected with preexisting
land tenure, forest management, and agricultural systems to

shape Madagascar's political economy and ecological landscape. Throughout this history, the precolonial, colonial, and postcolonial states have all blamed shifting cultivation, or tavy, for deforestation even as they have facilitated access to the eastern forests for commercial exploitation. Furthermore, state recognition of individual rights to cultivated land but not to forestland has created incentives for deforestation by commercial and subsistence farmers alike. During the colonial era, as commercial exploitation fueled the rapid destruction of classified forests, concerns about forest loss prompted the French colonial government to separate conservation and production areas into nature reserves, special reserves, and national parks for conservation and classified forests for the management of commercial wood supplies. These zones form the foundation of the current protected area and commercial forest networks. Nonetheless, the forest service has long been unable to enforce restrictions in these areas because of a lack of human, material, and financial capacity. Ultimately, in the face of enclosures of customary forestlands for commercial agriculture and conservation, combined with incentives to clear land to claim ownership, tavy has served not just as an agricultural practice and a way of honoring the ancestors but also as an important means to convert forest, which was owned by the state, into cultivated land to which the state recognized individual rights.

Neither foreign scientific nor foreign commercial interest in Madagascar's flora and fauna is a recent phenomenon. Madagascar was part of international trade networks for hundreds of years. Evidence of trade dates to as early as the tenth century, and archeological sites show evidence of numerous imported goods from the fifteenth and sixteenth centuries. Isle St. Marie became a well-established haven for pirates in the late seventeenth and into the eighteenth centuries, and Madagascar was a central component of eighteenth- and nineteenth-century Arab and Indian Ocean trade in forest products, rubber, gold, slaves, and tortoise shell (Campbell 1993; Dewar and Wright 1993; Dewar and Richard 2012; Walsh 2005). The precolonial Merina state granted rights to a variety of foreign traders to exploit forest products such as ebony, rosewood, and sandalwood as well as agricultural land (Campbell 2005; Evers, Campbell, and Lambek 2013).

As early as the mid-seventeenth century there are recorded attempts to classify Madagascar's species. Etienne de Flacourt, the governor of the French enclave of Fort Dauphin, wrote about the natural history of Madagascar in *L'Histoire de la Grande Île Madagascar* (The History of the Grand Island of Madagascar), first printed in 1659. His stories and others inspired a number of formal scientific explorations from Europe throughout the eighteenth and nineteenth centuries, during which time Europeans "discovered" a number of species. The British botanist James Petiver described the first species of lemur in 1703, and by the end of the nineteenth century the French naturalists Alfred Grandidier and Alphonse Milne-Edwards had pioneered the classification and study of almost all the European-known fauna and flora of Madagascar (Andriamialisoa and Langrand 2003). European discoveries in Madagascar were influential in scientific debates ranging from Darwinism to biogeography to human origins (Anderson 2013) as Madagascar was simultaneously portrayed as a "living Eden" and a "lost continent" (Feeley-Harnik 2001). These early expeditions and the scientific interest in Madagascar's species shaped subsequent research endeavors, classification efforts, and conservation approaches in the colonial and postcolonial eras.

Forest Policies Before Colonization

From the sixteenth to the eighteenth century, complex political systems developed in the central highlands and along the coast. The most well known of these was that of the Merina king Andrianampoinimerina, who ruled from 1787 to 1810 and who had defeated the surrounding kingdoms on the island by the time of his death in 1810. He used the corvée, or forced labor, to build dikes, water supplies, and foot roads so as to expand the frontiers of his domain. His son Radama I extended Merina authority over nearly the entire island. Radama I was followed by a line of kings and queens, including Queen Ranavalona I, King Radama II, Queen Ranavalona II, and Rainilaiarivony, who ruled the territory as prime minister under Queen Ranavalona III. The Merina kingdom had unified almost the entire island by the time of French colonization in 1896. Yet the Merina conquests and forced labor policies led to tensions between Merina and coastal people that continued into the colonial and postcolonial periods (Bertrand 2004; Brown 2000; Olson 1984).

There was no recorded forestry administration during the Merina Empire. However, numerous scholars cite as Madagascar's first forest regulations King Andrianampoinimerina's forest prohibitions, contained in the French translation of Merina oral histories *Tantaran'ny Andriana,* which was issued at the beginning of the nineteenth century (e.g., Kull 2004; Lavauden 1934; Olson 1984). These rules prohibited the gathering of fuelwood, tree cutting, and burning the forest except for forge work and charcoal making outside the forest. It is not clear if these rules were intended to apply to the entire island, just to royal forests, or just to the Merina-controlled highlands. Likely their primary purpose was to strengthen the king's power and to prevent rebellion against the state, as they forbade people to forge arms in the forest (Henkels 2001–2; Hufty and Muttenzer 2002; Kull 2004; Montagne and Ramamonjisoa 2006).

The next recorded forestry regulations appeared in the mid- to late nineteenth century. The Code of 101 Articles was issued in 1868, followed by the Code of 305 Articles (1881), which laid out comprehensive civil law, criminal law, and procedure. The Code of 305 Articles contained six forest-related articles, numbers 101–6, which established the first recorded comprehensive forest policy and forbade, with strict penalties, burning the forest, making charcoal or bamboo in the forest, cutting big trees, constructing houses in the forest interior, establishing new rice or maize fields in the forest, and cutting coastal forests.[2] Notably, the Code of 305 Articles reaffirmed state ownership of all forestland and recognized individual rights to cultivated lands (Keck, Sharma, and Gershon 1994; Leisz 1998). It also prohibited tree plantations, as tree planting could be used to claim land rights (Harper 2002; Henkels 2001–2; Kull 2004; Lavauden 1934; Olson 1984).

The French colonial forester Louis Lavauden (1934) later argued that the reason for this assemblage of regulations was likely that the Antananarivo-based government was concerned about the lack of sufficient fuelwood near the capital owing to the spread of tavy, wanted to keep people from hiding in the forest to avoid tax collection, and needed to protect the coast from European armies. Other scholars have similarly maintained that the forest policies were used to establish political and economic control over rural populations and valuable natural resources as well as to encourage the development of an economy based on permanent rice fields (e.g., Blanc-Pamard and Rakoto Ramiarantsoa 2003; Evers, Campbell, and Lambek 2013; Henkels 2001–2). It is also probable, however, that because the state did not have the wherewithal to enforce these regulations, local practice continued to govern most forest uses in areas far away from the capital (Montagne and Ramamonjisoa 2006; Ramanantsoavina 1973). Furthermore, as Campbell (2013) points out, Merina policies to promote economic development and imperial expansion through domestic industrialization likely caused much more deforestation in the highland and eastern woodlands than either foreign plantations or peasant cultivation.

Colonial Forest Policy and Land Politics

Both the British and French kept close trading and political ties to the Merina monarchy prior to French colonization. In 1890 the British recognized French authority over Madagascar in return for French acquiescence to British claims to Zanzibar. Five years later the French signed a treaty with the queen's representatives declaring a French protectorate over the entire island, and in 1896 the French parliament formally declared Madagascar a French colony. During the colonial period the French continued working closely with the Merina elite, which reinforced preexisting Merina–coastal tensions. As occurred in many European colonies, forestry agents trained in France and Germany brought tenets that were often ill-suited to the social and ecological characteristics of the forests in the colonies (Fairfax and Fortmann 1990; Montagne and Ramamonjisoa 2006; Peluso 1992). One such example, the eighteenth-century French forestry model, was based on utilitarian ideas of forest management for the greater good and government revenue, and it promoted the management of forests for rational economic exploitation by separating out areas for wood production and soil protection (Bertrand, Ribot, and Montagne 2004; Kull 2004).

At the start of the colonial period the French minister of agriculture sent three forest officers on a mission to Madagascar with the following goals: to become familiarized with Madagascar's forests and to understand its species; to manage their exploitation sustainably; and to secure control of future concessions (Lavauden 1934; Ramanantsoavina 1973). One of the first actions the colonial government took was to pass a *décret* ("decree") in 1900 that created the colonial forest service.[3] Then, in 1913, the state passed the first comprehensive colonial forestry codes. Based on the French National Forest law of 1854, the codes upheld that all forests belonged to the state and that all trees were state property (Lavauden 1934; Leisz 1998; Service de Colonisation 1913). The governor-general assumed authority over all forest concessions of 100 hectares or more (Harper 2002; Lavauden 1934; Ramanantsoavina 1973). Beginning in 1907 the French government also introduced eucalyptus, acacia, and pine in large plantations (Kull 1996), and it actively encouraged reforestation from 1910 to 1914 in the central plateau (Ramanantsoavina 1973). By 1928 it had established 1 million hectares of plantations (Olson 1984). These colonial practices mirrored preexisting struggles between the Merina elite and small farmers over forestland by securing it for commercial exploitation while banning tavy.

The colonial land tenure system reinforced the idea of individual rights to cultivated land by requiring a demonstration of a historical and continuing pattern of cultivation in order to obtain land title. However, most customary claims remained unrecognized during colonialism. A colonial law of 1896 stated that legal landownership could be established by "taking land" and could be held continuously as long as the land was cultivated. This law also accepted traditional claims to land as evidence of individual ownership and recognized cultivated lands, which were set aside to ensure local subsistence needs. Nonetheless, a decree of 1911 abrogated the use of traditional criteria for land rights in shifting cultivation areas. Another from 1926 stated that traditional testimonies could no longer be used as proof of rights to ancestral land and that, except for titled land, all land would be recognized as belonging to the state. In 1956 a decree recognized anew the right to use traditional testimony as proof of ownership of ancestral lands, but the *fokonolona*, or village-level administration, which regulated customary claims, was not allowed to register land officially (Hufty and Muttenzer 2002; Keck, Sharma, and Gershon 1994; Leisz 1998). In fact, most village lands are still not titled, and in many cases communal property rights, in which traditional authorities grant ownership to the person who clears the land, continue to the present (Healy and Ratsimbarison 1998; Keck, Sharma, and Gershon 1994; Leisz 1998; Schoonmaker-Freudenberger 1999). In this context, tavy remains an important means of staking claims to land, particularly in areas of contestation.

The French colonial state also used land privatization to encourage European settlement and to promote commercial export crops that could be used to repay debt that the precolonial state had incurred to France (Harper 2002; Keck, Sharma, and Gershon 1994; Leisz 1998). It offered liberal land grants to French metropolitan and Mascareen Creole settlers to establish cash crop plantations and to exploit mineral and forest resources (Campbell 2013); it favored French colonists by offering them 100 free hectares of public land while giving only 10 hectares to Malagasy who planted export crops (Brown 2000; Keck, Sharma, and Gershon 1994). In line with the colonial government's emphasis on the need to use land to own it, concession land had to be worked (Keck, Sharma, and Gershon 1994; Leisz 1998). Major colonial exports included ylang-ylang flowers, cloves, vanilla, and sugar cane in the north; cattle, rice, and maize in the west; and coffee on the east coast. Settler freehold was common along the eastern coastal region, where large plantations were established (Keck, Sharma, and Gershon 1994; Leisz 1998), and subsistence cultivators were forced increasingly from the east coast onto forested slopes. Concurrently, people from southern Madagascar moved north in search of wage work to pay taxes (Jarosz 1993). By 1934, as a result of increased production of cash crops, especially coffee, Madagascar had a trade surplus: coffee exports increased from six thousand tons in 1930 to forty-one thousand tons in 1938, and exports of vanilla, cloves, butter beans, graphite, and mica sustained the surplus until 1939. In the postwar period, sugar and sisal also became key export commodities (Brown 2000).

Major French companies secured concessions for mineral and forest exploitation, including for exotic hardwood species.

Forest area under concessions increased from 700 hectares in 1897 to 101,630 hectares in 1901, just five years after colonization (Bertrand 2004). By 1921 concession lands were estimated at 600,000 hectares (Olson 1984). Jarosz (1993) notes that forest concession owners were both European and Malagasy and that "Malagasy women formed a sizable proportion of those requesting concessions as a means of subsistence, while former military men demanded [concessions] as payment for service to the state" (374–75). Finally, the state created a system of taxes on rice fields, livestock, and markets designed to force subsistence farmers to produce export crops. It also used forced labor to develop infrastructure and public works, and companies could obtain unpaid labor from the state (Brown 2000; Harper 2002; Sodikoff 2005).

Confronted with enclosures of customary lands for commercial activities and forced labor policies, many rural people fled to the forests and survived as shifting cultivators. During this period and subsequent colonial repressions the forest became a place of hiding (Jarosz 1993). Throughout the colonial era the forests remained areas of protection from colonial forces: "From the standpoint of Malagasy people, the forests served as buffers against colonial penetration and the ensuing imposition of taxes and forced labor" (Sodikoff 2005, 414).

In 1942 the British took the island from the Vichy French through military action and then turned it over to the free French (Brown 2000; Kull 2004). In 1944 a major conference in Brazzaville to review the future of French colonies in Africa endorsed the assimilation of the colonies and their representation in a constituent assembly as well as in the subsequent French parliament, and in 1945 an elected representative council with budgetary powers and comprised of both Malagasy and French representatives was set up (Brown 2000). Nonetheless, Malagasy resistance to French policies was building, and on March 29, 1947, Malagasy nationalists revolted against the French. Joined by protesting villagers in the eastern rainforests, they burned and reappropriated land (Olson 1984). Their massacre of Europeans in the early days of the revolt set off violent retaliation from the French, and tens of thousands of Malagasy died in the reprisal (Brown 2000). Harper (2002) writes, "The rebellion symbolized the unity of the 'people of the forest' in their defiance of foreign control over their land and live-lihoods, and their victimization by these same policies and the op-pression which ensued" (89). These historical conflicts fostered distrust of foreign interests in forests that continues to pervade contemporary land politics in Madagascar.

Struggles over Tavy

The distrust was reinforced by the fact that the colonial govern-ment, even as it encouraged farmers to cut down trees in order to plant export crops and to establish plantations, continued to pass various regulations restricting tavy. Throughout much of Madagascar's history tavy was not only an agricultural practice comprised of shifting cultivation but also a means of connecting with and honoring the ancestors as well as a way to claim land (figs. 2.1, 2.2). While the specific customs vary by location and ethnicity, customary tenure often recognized that land belongs to the first person to clear it (Keck, Sharma, and Gershon 1994; Keller 2008; Leisz 1998). Where inheritance rights did not exist or had been disrupted, tavy was a method for claiming land that was recognized by other settlers and local officials (Ferguson et al. 2014). Once land was cleared, farmers could secure their continued property rights in the fallow season by planting banana trees, coffee plants, or other fruit trees (Keck, Sharma, and Gershon 1994; Leisz 1998). Customary tenure often integrated both individual holdings and commons in tavy areas, where water and forests were communal property and individuals could claim land as individual property by clearing it. In this regard, the part of the forest that touched the farmer's field was considered forest, and the farmer could control its use (Leisz 1998). Likewise, a piece of forest lying beyond al-ready cultivated arable land was thought of as land that was avail-able to future generations (Keller 2008). Given that precolonial, colonial, and postcolonial governments all recognized ownership of cultivated but not of uncultivated lands, and the state repeatedly claimed "underutilized land" and forests as its domain, it is unsur-prising that tavy became an important tool to convert forest—land owned by the state—into cultivated land to which the government recognized individual rights (Keck, Sharma, and Gershon 1994;

Figure 2.1. Hillside tavy, *Fandriana–Vondrozo Corridor. Photograph by Vokatry ny Ala.*

Figure 2.2. Cleared field, Fandriana–Vondrozo Corridor. Photograph by Vokatry ny Ala.

Leisz et al. 1995; Leisz 1998; Schoonmaker-Freudenberger 1999). Even as contemporary land reforms no longer consider untitled land as state property and recognize customary claims for the issuance of officially recognized private property rights, most rural communities continue to use customary property regimes (Evers et al. 2011; Ferguson et al. 2014; Neimark 2013).

From the onset of colonialism, governor-generals concerned about fire and deforestation ordered district forest officers to prohibit tavy in new zones, as opposed to previously cleared and regenerated sites (Kull 2004). Their efforts to contain peasant land use reflected regulations throughout French colonial Africa that were rooted in a long history of efforts to exclude pastoral use of forests and that prompted the so-called War of the Demoiselles in France in 1829–31 (Bertrand, Ribot, and Montagne 2004; Sahlins 1998). A decree of 1900 stated that trees of less than a meter in circumference could not be exploited (Lavauden 1934), and the forestry decree of 1913 banned tavy universally. The goal was twofold: to promote irrigated rice agriculture and to preserve the forest for commercial exploitation (Jarosz 1993; Oxby 1985). However, the impossibility of controlling tavy throughout the country meant that these orders were not always enforced (Kull 2004), and regional forest agents often tolerated tavy because it allowed them to extract taxes and enforce other unpopular state actions (Lavauden 1934).

Nevertheless, these tavy restrictions as well as earlier ones imposed by the Merina monarchy were perceived to oppose independence and to contradict land use traditions, and they elevated tavy to a symbol of resistance, where it became part of the Malagasy collective identity: "For the Malagasy peasant, shifting cultivation was a link to the ancestors, an affirmation of identity, a symbol and a means of resistance to state authority" (Jarosz 1993, 376). As an inherited practice and through the rituals of rice production, tavy has served as a means of connecting with and honoring the ancestors (Jarosz 1993). Where some land was redefined as nature reserves and other forests were exploited for industrial development, shifting cultivators connected tavy prohibitions with the enclosures. It became a means not only of claiming land and honoring the ancestors but also of resisting these enclosures (Harper 2002;

Jarosz 1993; Keck, Sharma, and Gershon 1994; Sodikoff 2005). That said, it is hard to separate the intertwined roles of tavy. As Kull (2002b) argues with respect to highland grassland fires in Madagascar, resistance is not simply peasant protest but "a multi-faceted livelihood-oriented strategy that grows out of the political-economic context as well as out of the contradictions inherent in state domination" (928). Like fires, tavy that is perceived as overt protest by policy makers may simply be peasants taking advantage of ambiguities in state legislation and enforcement.

Building again on French custom, and even as it prohibited rural Malagasy from engaging in tavy and commercial exploitation, the colonial state tolerated the use of forest products for noncommercial purposes as long as they did not interfere with commercial exploitation sanctioned by the colonial state. Lavauden (1934) wrote, "We find common traits in all the authors about the forest: that the forest is there for usage and not for individual appropriation, needing to be protected against abuse, particularly against fire, and at the same time, this notion of use, social assistance, on virtue of which the right to collect dead wood, which has in France always been considered a sort of right for the poor" (950). This adherence to a sort of moral economy (Neumann 1998; Scott 1976) also appears in the reforms of the forest service in 1913 and 1930, both of which recognized usage rights in the forest as long as they did not impede commercial exploitation. In a letter to the French president introducing the 1913 decree (Service de Colonisation 1913), J. Morel, the colonial minister, wrote, "Great care has been taken to respect and safeguard the traditional rights of the indigenous people, while ensuring that these rights do not become an obstacle to rational exploitation." While both decrees recognized use rights, they prohibited the selling of products taken under *droit d'usage* ("use rights"), claiming that such activities were not for personal use (Service de Colonisation 1913; Service des Forêts 1930). This colonial idea of allowing villagers use rights for noncommercial purposes only reappears in discussions about the new protected areas in Madagascar (see chapter 2) (figs. 2.3, 2.4).

Between 1900 and 1930 the forest service had limited ability to regulate concessionaires owing to a lack of human and financial capacity (Lavauden 1934; Montagne and Bertrand 2006; Montagne

Figure 2.3. Terraced lowland rice. Photograph by Desmond Fitz-Gibbon.

Figure 2.4. Roadside charcoal for sale. Photograph by Desmond Fitz-Gibbon.

and Ramamonjisoa 2006). It served primarily as a consultative body designed to provide policy guidance on sustainable forest use and agricultural development, and it was staffed by only a few Europeans (Ramanantsoavina 1973; Sodikoff 2005). The inability of the newly established forest service to control commercial exploitation in the face of the state's encouragement of concessions expansion led to massive forest loss, losses which scholars have estimated in slightly different ways. Jarosz (1993) reports that roughly 70 percent of the forest cover was destroyed between 1895 and 1925, and Boiteau (1982, cited in Hufty and Muttenzer 2002) reports that a third of the 12 million hectares of exploitable forests available at the time of French annexation was destroyed within fifty years. Finally, Lavauden (1934) cites a letter to France transmitting the 1913 forest decree, which states that more than 1.3 million hectares of forest disappeared during the early years of colonization.

Worry over this forest loss led French naturalists to push for a review of the forest service in the 1920s. Henri Perrier de la Bâthie, Henri Humbert, Alfred Grandidier, and others claimed that Madagascar had once been completely forested and that fire and tavy were the key causes of deforestation (Humbert 1933; Kull 2004). Grandidier and de la Bâthie asserted specifically that more than 200,000 acres of forest were lost each year (Bertrand and Randrianaivo 2003; Montagne and Ramamonjisoa 2006). Likewise, Humbert (1933) wrote, "In all the colonies, it is Madagascar that occupies the premier place. . . . It is also one of the countries in the world where the flora and fauna offer the most interest and are at the same time the most directly threatened by almost complete destruction" (212). Lavauden arrived in Antananarivo in the late 1920s to review and consolidate forest service texts and to reform the organization in order to redress forest loss (Kull 2002b; Ramanantsoavina 1973; Sodikoff 2005). The review resulted in two major decisions, both of which established the precedent for resolving the conflict between economic and ecological demands by geographically separating conservation and production zones.

The first outcome of the review was a decree issued in 1930 that reformed the forest service by tightening regulations and codifying the service's policing powers over both tavy and French concessionaires (Ramanantsoavina 1973; Saboureau 1958; Service des Forêts 1930). Specifically, it clarified forest boundaries, gave the governor-general more control over the permitting process, explicitly prohibited fires in and around the forest, reaffirmed concessionaires' responsibility for transgressions, encouraged reforestation, and gave forest service agents stronger enforcement powers, including the authority to determine allowable indigenous usage rights (Kull 2004; Lavauden 1934; Ramanantsoavina 1973). In particular, Lavauden proposed that the central forest service assume greater control over concessions. The move to centralize the permitting process was intended to regulate French concessionaires who were exploiting ebony, industrial wood, charcoal, and bark for textile fibers (Harper 2002; Sodikoff 2005). Unfortunately, these reforms had little impact on forest loss. Despite its increasingly repressive policies, the service continued to lack regional and local staff and thus the capacity to enforce regulations against either commercial exploiters or shifting cultivators. As a result, it had little legitimacy with either group (Kull 2004; Montagne and Bertrand 2006).

At the same time, there was growing economic pressure from abroad, particularly to furnish supplies and labor to France for and between the two world wars. Madagascar became a key supplier of timber, rubber, fibers, and personnel. During the interwar period the state commenced a public works campaign to construct roads, railways, bridges, ports, and buildings, all of which demanded wood for their construction; the increasing use of rail transport gave rise to a demand for fuelwood and triggered the exploitation of timber from regions that previously were difficult to reach (Campbell 2013; Jarosz 1993; Ramanantsoavina 1973; Sodikoff 2005). During World War II the forest service was charged with developing wood production, charcoal, rubber, and other products needed both for domestic economic growth and to furnish the Allies with materials (Saboureau 1958). After the war it continued to manage resources for domestic and foreign economic needs, including fibers, textiles, rubber, raphia, and fuelwood (Ramanantsoavina 1973).

In the postwar era the state began rapidly securing forestland for both commercial wood use and conservation purposes. Orders in 1955 and 1957 introduced the idea of *forêts classées*, or "classified forests," as reserves to ensure a consistent supply of commercial wood for use by the state. The surface area covered by classified forests increased from 1,315,000 to about 3,250,000 hectares by the end of 1959 (Ramanantsoavina 1973). The forest service allowed commercial and traditional harvesting in classified and state forests by permit, and, reflecting the belief in protecting noncommercial use rights, local residents were allowed to practice traditional harvesting of forest products, including firewood, house-building materials, fibrous plants used in basket and mat weaving, and medicinal plants. Again, however, lack of management capacity meant that the forest service did not enforce restrictions in these areas against either shifting cultivators or concessionaires.

Protecting Madagascar's Flora and Fauna

The second outcome of the forest sector review of the 1920s was the establishment in 1927 of ten nature reserves covering approximately 350,000 acres (Décret 1928; Saboureau 1958). Pressure from foreign scientists to protect Madagascar's flora and fauna had begun growing during the early years of colonization. In 1902, two years after the forest service was established, Governor-General Joseph-Simon Gallieni created the Académie Malgache (Malagasy Academy) with the aim of "merging in one structure all research efforts on the island conducted by French and Malagasy scientists (many of whom were trained in France), as well as those of other nationalities" (Andriamialisoa and Langrand 2003, 4). Common research effort reached its peak in 1929 with the launch of the Mission Zoologique Franco-Anglo-Américaine (Anglo-American-French Zoology Mission), which collected an enormous quantity of specimens (Andriamialisoa and Langrand 2003).

After the forest service review, the Malagasy Academy and the forest service proposed the creation of thirty nature reserves that would be "equivalent to the forest reserves, but with a completely different purpose" (Humbert 1933, 212) and that would be closed completely to exploitation as well as to tourism. The reserves had three primary goals: to ensure conservation of representative types of vegetation; to provide scientific laboratories for the study of species; and to ensure the protection of areas important to climate regulation and protection of water sources (Humbert 1933; Saboureau 1958). In the end only ten were accepted, to be managed by the forest service (Andriamampianina 1987; Décret 1928; Kull 1996; Ramanantsoavina 1973).[4] The sites ultimately chosen were far from major population centers and thus not practically exploitable for commercial wood supplies (Andriamampianina 1987; Saboureau 1958). All usage rights, including hunting, fishing, and harvesting by local communities, were prohibited, and scientific research was allowed only with the permission of the Natural History Museum of Paris (Andriamampianina 1987; Décret 1928; Randrianandianina et al. 2003).

In the last years of French rule and the early years of Malagasy independence the state established additional categories of conservation areas, including national parks and special reserves.[5] Decrees issued between 1950 and 1955 charged the forest service with classifying and managing these new conservation zones. The goal of special reserves was to set aside, under slightly more lenient regulation, sites with particular scientific, touristic, climatic, or economic characteristics or that were important water sources for hydroelectric energy but that did not meet the requirements of strict nature reserves or national parks (Andriamampianina 1987; Saboureau 1958). Eight special reserves were established in 1956 and seven in 1958. By 1965 twenty special reserves had been created, the majority of which were areas proposed but not accepted as strict nature reserves in 1927 (Andriamampianina 1987).

Like the advocates of rational forest exploitation, those who promoted conservation reserves drew on internationally circulating ideas, in this case preservationist notions, which began gaining popularity toward the end of the nineteenth and into the early twentieth century. The exclusionary approach embodied in Yellowstone Park became an international model for national parks and reserves. In 1892 the South African president Paul Kruger established the Sabi Game Reserve, now the Kruger National Park, and in 1925 King Albert of Belgium set up a gorilla sanctuary, which became the Albert National Park and then the Virunga National Park in the Belgian Congo (Oates 1999).

While many of the first national parks in continental Africa were reclassified as game reserves originally designated for hunting,

Madagascar's parks emerged out of a history of scientific interest in the country's flora and fauna. A key catalyst of Madagascar's reserves was the National Committee for the Protection of Colonial Fauna, instituted in Paris in 1923, which concluded in 1925 that the French government should establish national parks in all French colonies (Andriamampianina 1987; Fenn 2003; Humbert 1933; Ramanantsoavina 1973). A number of authors have suggested that Madagascar's early reserves inspired the decision of the London Convention on the Preservation of Fauna and Flora in 1933 to create a new category of strict nature reserves for the conservation of flora and fauna species "not sufficiently protected" by existing reserves (Convention Relative to the Preservation of Fauna and Flora in Their Natural State 1933; see also Andriamampianina 1987; Heijnsbergen 1997; Ramanantsoavina 1973; Saboureau 1958).

In the 1960s Africa became a focus for expanding international conservation efforts (Adams and Hulme 2001), reinvigorating the momentum of the twenties to protect Madagascar's species. In 1961 the Arusha Declaration on Conservation stressed both wildlife conservation and sustainable resource development, and the IUCN and United Nations Food and Agriculture Organization (FAO) initiated the African Special Project, which entailed IUCN missions to seventeen African countries. These resulted in a new African Convention on the Conservation of Nature and Natural Resources that was adopted by the Organization of Africa Unity in 1968. The convention incorporated a more utilitarian approach to conservation than some of the efforts in the 1930s, specifically designating conservation areas as "any protected natural resource areas, whether it be a strict natural reserve, a national park or a special reserve" and underscoring that "the utilization of the natural resource must aim at satisfying the needs of man according to the carrying capacity of the environment" (African Convention 1968). The IUCN, the National Parks Commission (now the World Commission on Protected Areas), and other partners convened the First World Conference on National Parks in 1962 (Chape et al. 2005), and there was a special conference on African conservation problems in 1963 in Bukavu, Belgian Congo (Adams and Hulme 2001).

In Madagascar two decrees established Amber Mountain and Isalo National Parks in 1958 and 1962, respectively. These decrees cite the London Convention (Président de la République 1962; Président du Gouvernement 1958), and, in line with the London convention's guidelines, the goal of the parks was both to protect species and habitats and to set up tourist sites in order to stimulate greater public support for conservation. Unlike the initial nature reserves, the two parks allowed tourism and limited use by local residents (Andriamampianina 1987; Kull 1996; Président de la République 1962). Various other natural and historical sites and zoological reserves were also classified during these years (for more information, see Griveaud and Albignac 1972; Kull 1996; Saboureau 1958), and in 1960 the government of Madagascar put in place strict rules against fishing, hunting, deforestation, and bush fires. Again, however, enforcement of regulations was minimal given the difficulty in reaching rural areas and lack of transportation for agents (Montagne and Bertrand 2006; Ramanantsoavina 1973). Collectively, these three categories of protected areas comprised the national parks system until the initiative to expand protected areas in the early twenty-first century.

Postcolonial Land Use Incentives

On June 26, 1960, Madagascar officially gained independence from France, and Philibert Tsiranana won the presidential election conducted under French auspices (Marcus 2004). The independent government kept much of the colonial system of land tenure during the First Republic (1960–72). The state continued to regard all land not individually registered as state land; it claimed rights to all forests and trees except those on titled land; and it reclaimed underutilized private land (Healy and Ratsimbarison 1998; Leisz 1998). It also revived reforestation efforts aimed at ensuring fuelwood and construction supplies (Montagne and Bertrand 2006). A national reforestation campaign in 1963 aimed to plant 7,000 hectares a year of new forests on private land. Each citizen was required to plant one hundred seedlings a year or pay a fine of one hundred francs, although in 1971 this mandate was replaced by a reforestation head tax (Gade and Perkins-Belgram 1986; Kull 1996; Olson 1984), which brought in eighty to one hundred million francs a year, or the equivalent of three to four times the annual forest service budget

(Olson 1984). Many people associated this policy with colonialism, and in 1972 it was abolished, although it was occasionally revived in subsequent years. A reforestation campaign was a key priority of the NEAP, for example, although according to Madagascar government officials it was one of the most underfunded priorities in the plan because of donor disinterest.[6]

The constitution of 1960 reiterated an individual right to property under the condition that the individual exploit the land as part of the greater public interest. When lands of more than 5 hectares were not used for more than five years, the constitution authorized their transfer to the state. Individuals unable to use the land gave up their rights to it, and the land was free to be transferred to a more productive owner / user. In this manner land tenure laws again emphasized the social obligation to develop land. While the political regime changed in 1972, the land law stayed virtually the same. The constitution of 1975 guaranteed individual property rights for material well-being but stated that if individual property rights ran counter to social needs the state could limit them (Keck, Sharma, and Gershon 1994; Kull 1996). Thus after independence, as in the colonial period, there were strong incentives to convert forests into cultivated land.

These incentives, combined with the legacy of enclosures and the lack of enforcement capacity, again fueled deforestation by farmers and concessionaires. The First Republic is remembered as a time when strict authorizations existed but were not necessarily followed (Kull 2004). The replacing of traditional forest tenure, management norms, and regulations with the idea that forests belonged to the government "would later result in the degradation of these areas as forestry laws were no longer enforced and the *fanza-kana* [government] gradually lost the respect of much of the Malagasy population" (Fenn 2003, 1497). Sodikoff (2005) argues that, in more contemporary times, "tavy is in part a refusal to abandon the measure of liberty gained in the decades after French colonialism when the rules grew softer and more malleable" (429).

The Second Republic, which came to power in 1975, also maintained state control over land and resources, and, again, numerous scholars and policy makers refer to the socialist period after the second independence as one in which environmental degradation increased substantially. The revolution encouraged people to clear land for food production in order to address the poverty that had resulted from economic recession in the early 1970s. Furthermore, the state encouraged rural land titling by deeding land to rural collectives and loosening controls over natural resource use, which helped increase President Didier Ratsiraka's influence in rural areas (Simsik 2002). These policies coincided with diminishing resources allocated to the forest service and its resulting inability to enforce forestry laws, creating incentives for forest clearing (Blanc-Pamard, Pinton, and Rakoto Ramiarantsoa 2012). Forest service budgets, which peaked in 1960 and were stable for a decade, began declining in the 1970s and 1980s. This led to the cessation of personnel recruitment and public planting programs as well as to reduced control over forest districts (Montagne and Ramamonjisoa 2006). The result was accelerated deforestation throughout the rural areas: "Rural Malagasy interpreted these two events to mean that they were free to exploit natural resources as they wished, and the immediate and more visible result was that natural resource conditions degraded at an alarming rate" (Simsik 2002, 234). Moreover, as the standard of living decreased and controls on rice production led to the loss of Madagascar's self-sufficiency in rice production, the country began facing an economic crisis (Hufty and Muttenzer 2002).

Continuing the commitment to the separation of conservation and production zones, the postcolonial forest service emphasized rational use of natural resources, specifically through agricultural production, reforestation, wood use, and forest exploitation (Ramanantsoavina 1973). As the former head of the forest service George Ramanantsoavina (1973) wrote, "Without abandoning its traditional role of protection and conservation, the Forest Service has oriented itself presently toward the more dynamic concepts of development" (34), while "address[ing] these problems of the environment with the system of protection already well known of classifying national parks and reserves to maintain characteristic ecosystems" (32). In this way the state addressed the ongoing conflict between production and conservation demands by spatially separating the two, isolating "the environment" in nature reserves, special reserves, and national parks rather than developing capacity to sustainably manage commercial extraction. This policy of separation continued into contemporary times.

Land Tenure, Commercial Forestry, and Conservation Politics

This overview of the history of the intertwined land tenure, forestry, and conservation politics in Madagascar offers a glimpse into the ideological and material struggles that have characterized the country's eastern landscape for hundreds of years. Throughout this history foreigners brought demands for Madagascar's forest products as well as ideas that shaped their management. State-subsidized commercial agriculture and timber exploitation, combined with state recognition of individual rights to cultivated land but not to forest, created incentives for deforestation. Furthermore, in the face of state prohibition of tavy, repression of peasants, and enclosures of customary forest for expanding commercial agriculture and conservation, tavy became a multifaceted peasant strategy, serving as a means of livelihood, a way of connecting with the ancestors, and a tool to claim land.

As rural peasants, the Madagascar state, commercial interests, and the international conservation community endeavored to claim Madagascar's eastern rainforests for science, economic gain, and livelihoods, they laid the foundation on which later environmental politics were built. The political, economic, and cultural relationships that developed among traders, politicians, and citizens as well as their ideas about nature and economy established the foundation for contemporary approaches to using and saving Madagascar's resources.

As I explore in the following chapters, contemporary conservation efforts are replaying historical struggles over Madagascar's forests. By failing to recognize the intertwined nature of foreign investment, land tenure politics, and rural livelihoods, the initiative to expand Madagascar's protected areas has reinforced, rather than redressed, the drivers of deforestation in Madagascar. In particular, it has reclaimed control of Madagascar's forests, restricting peasants' commercial activities. Yet the state lacks the capacity and incentive to arrest expanded mineral and timber exploitation and large-scale agricultural expansion. The escalating struggle over Madagascar's forests among traders, conservationists, the state, and rural peasants has reinforced incentives for claiming land by clearing it.

Importantly, however, current conservation efforts are being implemented under conditions of changing power dynamics among public, private, and nonprofit entities. Until the mid-1980s forest policy in Madagascar was primarily a project of rulers, even as the state was not unified, autonomous, or a coherent entity and was always influenced by foreigners. In the mid- to late 1980s rising foreign interest in protecting Madagascar's biodiversity converged with donors' neoliberal push for a reduced state, a liberalized economy, and civil society and private sector participation in state policy formulation and implementation. The result was increasing international influence—by both conservationists and commercial exporters—over formerly state policy-making processes. The next three chapters move across the United States, England, and Madagascar to trace this emerging influence.

CHAPTER THREE

Setting the Biodiversity Conservation Stage

There is no question that [the Madagascar environment program] was initially driven by the conservation community. . . . In the mid-1980s, there was a growing global awareness of the importance of biodiversity. Then the World Bank created the National Environmental Action Plans in response to the global movement, and then the conservation organizations took advantage of this attention and used it as an opportunity to build an environmental program. The U.S. really got on the bandwagon at that time. It created the [USAID] global environmental office (in Washington, D.C.). There was a convergence of money from Congress, and then we had the Rio Conference, so U.S. funding was growing.

—INTERVIEW WITH A USAID MADAGASCAR OFFICIAL, November 24, 2005

IN THE 1970s A group of foreign and Malagasy scientists, many of whom were well-known primatologists, began a campaign to build global awareness of the plight of Madagascar's flora and fauna. Through a series of meetings, conferences, and trips these advocates developed relationships with policy makers, outlined conservation priorities, and drafted institutional protocols that continue to influence Madagascar conservation politics today. When the World Bank staff began looking for countries in which to launch the bank's new environmental agenda, Madagascar offered the perfect place given that scientists, conservationists, and policy makers had already established solid collaborations and identified key priorities there.

At the forefront of a World Bank initiative that required borrowing countries to produce NEAPs in order to qualify for structural adjustment lending, the three-phase, fifteen-year Madagascar NEAP, coordinated and funded by donors and the government, aimed to mainstream environmental policies into the country's overall development plans. It became a model for other countries putting together national environmental plans. In line with the rise of participatory development and the endorsement of civil society involvement in foreign aid policy at that time, the World Bank recruited scientists, NGOs, and consultants to help develop the NEAP's priorities and implement its programs. As members of the steering committees that guided the NEAP, non-Madagascar state actors actively shaped the NEAP's priorities throughout its three phases (Duffy 2006; Kull 1996; Sarrasin 2007b).

Building on the previous chapter's analysis of the historical conflicts over Madagascar's forests, in this chapter I examine how a confluence of political and economic forces in the mid-1980s gave scientific advocates, many of whom were Americans or associated with American institutions, an opportunity to transform their campaign to protect Madagascar's flora and fauna into a well-funded foreign aid agenda. Having a decidedly American-centric perspective, this chapter offers not a complete account of the history of Madagascar's environmental program but a view of some of the historically grounded relationships that comprise it. My approach elides the critical influence of other multilateral and bilateral donors as well as that of the Madagascar government. Likewise, even as I have focused on personal relationships, because I agreed to protect individual confidentiality except in selected cases, I have downplayed the role of specific individuals, many of whom continue to have an important influence on Madagascar conservation politics. Even though some of the organizations and people discussed in this chapter are not American, the locations in which many of these events took place offered critical access to Washington, D.C.–based policy makers, including the World Bank and USAID, which together eventually funded the bulk of the biodiversity program. Thus these events constituted nodes of social relationships (Massey 1999) that brought together and aligned diverse actors toward a common pursuit.

The Seeds of Foreign-Funded Conservation

Throughout the colonial era European scientists continued conducting research on Madagascar's flora and fauna (see chapter 2). From the end of World War I to the late 1960s, the Académie Malgache and the Institut de Recherche Scientifique de Madagascar (IRSM) (Institute of Scientific Research in Madagascar) facilitated numerous zoological expeditions, and foreign scientific interest in Madagascar's lemurs intensified in the mid-twentieth century. Supported by IRSM, the French primatologists Jean-Jacques Petter and Arlette Petter-Rousseaux started studying the lemurs in the 1950s. In 1960, the year of independence, the broadcaster and naturalist David Attenborough, aided by the ornithologist Georges Randrianasolo of IRSM, made the first commercial film about wild lemurs for a Western audience. Collaboration among Malagasy and foreign researchers interested in Madagascar's flora and fauna also led to various scientific research projects under the Centre National de la Recherche Scientifique (National Center for Scientific Research). In the 1960s the American anthropologist John Buettner-Janusch brought lemurs back to Yale University to study and later founded the Duke Primate Center (now the Duke Lemur Center) (Jolly 2015) (figs. 3.1, 3.2). Encouraged by Buettner-Janusch, the primatologist Alison Jolly began studying ring-tailed lemurs in 1962, followed by Robert Martin, Alison Richard, and Bob Sussman, who began their research in the 1970s. They all attended the Conférence

Figure 3.1. *Duke Lemur Center. Photograph by Catherine Corson.*

Internationale sur la Conservation de la Nature et de ses Ressources à Madagascar (International Conference on the Conservation of Nature and Resources in Madagascar) in 1970 (Andriamialisoa and Langrand 2003; Jolly and Sussman 2007).

Figure 3.2. Coquerel's Sifaka at the Duke Lemur Center. Photograph by Catherine Corson.

The conference laid the foundation for the subsequent rise of foreign interest in Madagascar's environment. Sponsored by a number of international research and conservation organizations, it was organized by Petter of the French Musée National d'Histoire Naturelle (National Museum of Natural History) and Monique Ramanantsoa Pariente, the daughter of General Ramanantsoa, who became Madagascar's interim president in 1972.[1] "The idea of organizing a conference," a former Malagasy official recalled, "came from a few foreign scientists, and some Malagasy, who were worried about the growing degradation of forests."[2] It focused primarily on nature conservation: the slogan "Malagasy Nature, World Heritage" was visible everywhere (Jolly and Sussman 2007, 28), and the keynote speakers, Calvin Tsiebo, the vice president of Madagascar, and H. J. Coolidge, the president of the IUCN, underscored the endangered status of the country's flora and fauna. The conference reflected the perception of Madagascar's special place in an expanding international conservation movement, a legacy summarized in the conference proceedings:

> Since the international movement to promote nature conservation took off world-wide, the protection of Malagasy nature has always been an object of great attention. The reports of the International Conference on the Protection of Nature held in 1947, in Brunnen, Switzerland, where the IUCN was founded, show that the special case of Madagascar already held substantial attention at this time. This international scientific interest appeared already in even older testimonies: in a series of measures that Madagascar took, in particular the 1927 creation of a network of integral reserves which form, until this day, the most important protected areas in the country. (IUCN 1972, 5, my translation)

The attendees, primarily foreign and Malagasy researchers and conservation NGO representatives, produced a variety of recommendations and resolutions for action. These encompassed seven subthemes, including: the international scientific importance of Madagascar; soil, climate, and vegetation; freshwater and marine issues; fauna; parks; agroforestry; and nature conservation (IUCN 1972).[3]

The tension between conservation and development that was to permeate the country's subsequent environmental agenda surfaced at this conference. Kull (1996) cites an intervention by Etienne Rakotomaria, the director of scientific research, critiquing foreign organizations and scientists for dominating the discussions. Likewise, Jolly recalls being escorted out of the conference by Charles Lindbergh and Sir Peter Scott, the founder of World Wide Fund for Nature (WWF-International), after presenting a paper she and her husband, the well-known economist Richard Jolly, had written entitled "Conservation: Who Benefits and Who Pays?" Lindbergh and Scott "instructed her that although it was obvious that poor people who lose their land pay most of the price of reserves, she should not say so. It would set back the cause of conservation to raise such issues." She contrasts their reaction to those of Perez Olindo of the Kenyan Game Department and David Wasawo, the vice chancellor of the University of Dar es Salaam: "'High time someone said that!' they declared. 'Come and stay with our families in Kenya!'" (Jolly and Sussman 2007, 28). As she recollects further, "That paper did not appear in the published proceedings. I did a very brief paper that I scribbled at the time because someone said 'do tell us about lemurs.' That made it into the proceedings. But *Who Benefits and Who Pays?* did not." She went on to explain the reasons behind this effort to silence her: "The conservationists had been fighting a battle to get heard, particularly in Africa. So the last thing they wanted was something that raised a question that threatening."[4] The tension between those advocating a pure conservation agenda and those pushing integrated conservation and development continues to this day, as do efforts to downplay public discussion of the potential negative social impacts of conservation.

While the conference provoked increased awareness of Madagascar's environment, it was ill-timed and did not generate an influx of foreign aid for environmental issues as the Madagascar authorities had hoped. Instead, the momentum that it inspired stalled in the wake of the revolution that followed. Concerns expressed by the conference participants about foreign interests driving conservation represented a spreading dissatisfaction with the degree to which the French continued to influence political and economic affairs in Madagascar. At independence, foreigners controlled 80 percent of Madagascar's economy, and the French held many key government and university positions (Brown 2000; Kull 1996). A student-led general strike against economic conditions in 1972, primarily in objection to the French domination of university, schools, and government, shut down much of the country. President Tsiranana declared a state of emergency, dissolved parliament, and turned over power to Gen. Gabriel Ramanantsoa. The revolution of May 1972 is often referred to as the second independence from French domination. Col. Richard Ratsimandrava took over power on February 5, 1975. When he was assassinated, Lt. Comm. Didier Ratsiraka assumed power. A referendum on Ratsiraka's presidency and a call for a new socialist government held on December 21, 1975, led to the Second Republic, with its Leninist scientific socialism agenda and an emphasis on poverty reduction.

During the Second Republic the government turned away from France and other Western countries and toward the Soviet Union, North Korea, and China. It implemented agrarian collectivism, nationalized key sectors of the economy such as agriculture, and borrowed heavily from external sources to finance a national investment plan (Marcus 2004; Sarrasin 2005; Sodikoff 2007). From the launch of the nationalist agenda in 1972 until the mid-1980s, when the government turned to the West for foreign aid, many foreigners found it difficult to visit the island, and the Ratsiraka government gave only a handful of biologists research permits during this time (Andriamialisoa and Langrand 2003; Fenn 2003; Jolly and Sussman 2007). Foreign scientists who did enter often came through higher education system partnerships.[5] Given Madagascar's socialist bent, Western governments and NGOs alike were reluctant to pledge significant funds to conservation there.[6]

Nevertheless, the seeds of the foreign-funded conservation program of the future were being planted. One of the 1970 conference resolutions called for the creation of a WWF-International office in Madagascar, which was established by presidential decree in 1979. The resolution mandated that the director be Malagasy and work closely with governmental authorities (Repoblikan'i Madagasikara 1979). At the behest of Petter, Barthélémi Vaohita was appointed the WWF-International representative, and he became very influential in the eventual development of the NEAP. An accord between WWF and the Madagascar Ministère de l'Enseignement Supérieur et de la Recherche Scientifique (MESupReS) (Ministry of Higher Education and Scientific Research) then established the WWF program of action in Madagascar. It acknowledged the need for information about both park management and ecosystem dynamics; recommended preparation of inventory of fauna and flora; obligated the state to coordinate its agencies; mandated that all research projects and resulting publications be undertaken in a collaborative manner; and committed WWF to mobilizing foreign aid for conservation (MESupReS and WWF-International 1983).

These initiatives and others by WWF formed the basis of the ensuing conservation agenda. In 1975 Guy Ramanantsoa of the University of Madagascar, Richard, and Sussman created the Bezà Mahafaly reserve as a training ground for students at the University of Madagascar's School of Agronomy (Richard and Ratsirarson 2013). A project leader attributed the idea for its bottom-up approach to "the vision of agronomist Gilbert Ravelojaona, who said that the problem with Madagascar's protected areas system was that it was top down and imposed by the French."[7] With an initial sum of US$120,500 for seven years from WWF-U.S., the site became an official special reserve in 1986 and thereafter one of the first USAID-supported environmental projects in Madagascar. In a concurrent effort to raise public awareness about Madagascar's flora and fauna in the United States, Thomas Lovejoy of WWF-U.S. commissioned Jolly to write *A World Like Our Own*, which was published in 1980 (Jolly and Sussman 2007). Despite these initiatives, the perception that Madagascar's environment was in grave danger increasingly concerned scientists and conservationists, and in the late 1970s and 1980s they began organizing.

A series of meetings, trips, and conferences, some of which took place outside of Madagascar, cemented critical relationships among Malagasy government officials and a group of scientists. In 1979 Césaire Rabenoro, the president of the Académie Malgache, hosted an international meeting on lemur biology. Gerald and Lee Durrell of Jersey Wildlife Preservation Trust (JWPT) (now the Durrell

Wildlife Conservation Trust), among others, attended this meeting (Jolly and Sussman 2007). In November 1981, following a visit by the World Wide Fund for Nature Madagascar (WWF-Madagascar) representative Vaohita to the United Kingdom, a group of foreign scientists working in Madagascar held an informal gathering in Cambridge, England, to discuss how to promote nature conservation in Madagascar. They decided to invite the relevant Malagasy authorities to a follow-up meeting. In February 1983 the JWPT hosted the follow-up meeting on the island of Jersey in the Channel Islands (Durrell 1983), where Gerald Durrell had founded the Jersey Zoological Park in 1959 as a breeding center for endangered species. Seven participants in the Cambridge meeting, Malagasy representatives from the relevant ministries and technical organizations, and additional representatives of various universities, museums, and wildlife organizations based in the United States, the United Kingdom, and mainland Europe attended the Jersey meeting. The goal was to highlight foreign interest in Madagascar's flora and fauna for the Malagasy authorities and to address the problematic process for obtaining research permits. It was at this meeting that Petter raised the idea of holding a follow-up conference to the 1970 conference (Durrell 1983).

At this meeting also Madame Berthe Rakotosamimanana, then the director of scientific research with MESupReS, laid out a plan to facilitate the permitting process for foreign scientists. While foreign scientists were concerned about the challenges of obtaining research permits, the Madagascar government was overwhelmed by the number of uncoordinated proposals from foreigners wanting to conduct scientific research in the country. Although some research institutions, such as Strasbourg and Duke universities, had formal agreements with MESupReS, individual researchers often approached the Madagascar government separately. In an attempt to address the issue, WWF-International and the ministry signed an annex to their existing accord that established an International Advisory Group of Scientists (IAGS) to coordinate biological research conducted by foreigners in Madagascar (MESupReS and WWF-International 1983). Composed of Roland Albignac, Lee Durrell, Jolly, Bernd-Ulrich Meyburg, Petter, Peter Raven, and Richard, this group screened biological research proposals, which they then forwarded to WWF and the appropriate ministries in Madagascar, with the goal of expediting permission to conduct research. Reflecting the priorities of the WWF-International program and the interests of the group, the IAGS emphasized the need for biological surveys: "For conservation purposes, the research most urgently needed by Madagascar concerns up-to-date biotic inventories of her last remaining natural habitats" (Durrell 1984). Martin Nicoll, Sheila O'Conner, and Olivier Langrand began conducting research in the early 1980s, and in 1986 WWF-International hired Nicoll and Langrand to conduct a review of existing protected areas and to propose new priority areas to protect these habitats.[8]

These meetings cemented critical personal relationships, introduced institutional protocols, identified programmatic priorities, and institutionalized the influence of the scientific community on Madagascar environmental politics, all of which would shape the country's future conservation agenda. In particular, the emphasis on biological surveys of populations continued as the environmental program expanded, and as the group channeled funds and permits toward specific research priorities these early assessments created the scientific basis for the biodiversity portion of the NEAP and the foundation for the eventual expansion of Madagascar's protected areas. As the country reopened to the West after the decline of the socialist regime in the 1980s, these scientists and policy makers found themselves at the center of a global political transformation.

The International Discovery of Madagascar

The 1980s marked an important turning point in Madagascar's environmental history. As a result of extensive borrowing and capital flight, the Madagascar government had to abandon much of its socialist agenda, and it began turning to the West. Its budget deficit was over 40 percent; exports comprised less than 50 percent of imports; industry faced a shortage of raw materials and spare parts; inflation was at 27 percent; and foreign debt was over US$1 billion. The government signed its first IMF agreement in 1980, under which donors agreed to reschedule or refinance Madagascar's debt in exchange for the acceptance of an IMF stabilization program.

The program entailed currency devaluation, fiscal austerity, and liberalization of the economy. The resulting restructuring of the Madagascar state and economy reduced the role of central government institutions, promoted private sector and NGO involvement in formerly state policy, and opened the economy to foreign investors. As it turned to the West, donor assistance rose rapidly (Brown 2000; Kull 1996; Marcus 2004). Bilateral donor assistance increased from US$36.3 million in 1976 to US$217.6 million in 1988 and to US$365.5 million in 1991 (Horning 2008a).

At this critical historical juncture Madagascar burst into the international limelight. Key events underpinned its emerging international fame. These included the discovery in 1986 of the golden bamboo lemur, the growing awareness of the rosy periwinkle's use as a treatment for childhood leukemia, and a widely publicized satellite image of Madagascar in 1984 from the American space shuttle *Discovery*, which showed "Madagascar bleeding to death" as reddish-brown water from eroded soils poured into the Betsiboka River estuary off the northwest coast (Gezon 2000; Simsik 2002). As Kull (1996) writes, "Madagascar returned to the world map after a decade of isolation largely through the lens of conservation—perhaps literally through the camera lens, as images and stories of lemurs, chameleons, orchids, erosion, and deforestation made it to television documentaries and popular publications" (67). A senior international conservation NGO representative who had been working in Madagascar at the time recalled, "What really happened is that all of a sudden at the national level, at an international level, beyond the circle of scientists, there was a discovery of the importance of biodiversity in Madagascar."[9]

Madagascar's fame was fueled by rising global interest in biodiversity and sustainable development. Sponsored by the Smithsonian Institution and the National Academy of Sciences, the 1986 National Forum on BioDiversity was convened in Washington, D.C., with the explicit intention of raising congressional awareness of global species loss (Takacs 1996). At the forum Russ Mittermeier proposed the idea of mega-diverse countries and identified Madagascar as one of the top six most important countries (Mittermeier 1988; see also Mittermeier et al. 1998). In 1988 the ecologist Norman Myers similarly proposed the idea of protecting critical regions with high concentrations of endemic species that faced habitat loss and proclaimed Madagascar one of the world's top ten biodiversity hotspots (Myers 1988, 1990; Myers et al. 2000). As a result, much of the rising global interest in biodiversity focused on Madagascar as a high-priority country.

At the same time, environmentalists were pushing development donors to fund environmental programs under the auspices of sustainable development. First articulated in 1980 in the IUCN World Conservation Strategy, the popular concept of sustainable development offered a way for aid donors to endorse environmental issues without opposing economic growth. The IUCN strategy also recommended that countries prepare national conservation strategies (IUCN 1980), and in 1984 Madagascar became the first country in the Afro-tropics to follow the IUCN recommendation. The country's strategy highlighted many of the same issues raised at the 1970 conference, and it put forth specific recommendations related to forests, grazing land, agriculture, parks, soil erosion, fishing, and tourism (Repoblika Demokratika Malagasy 1984).

Like the 1970 conference in its focus on the preservation of Madagascar's flora and fauna, the 1984 strategy reflected the IUCN's framing of conservation as a means of advancing, rather than an impediment to, sustainable development. It linked natural resource management to food security: "It appears more and more obvious that the management of natural resources for sustainable development is an urgent necessity and should constitute the pivot around which the government policy to secure food self-sufficiency will hinge in the future" (Repoblika Demokratika Malagasy 1984, summary, my translation). In doing so, it marked a transition in emphasis from nature conservation, the focus of the 1970 conference, to the environment,[10] and it established the groundwork for a national environmental agenda that was broader than biodiversity conservation. In 1984 Vaohita convinced every Malagasy minister to sign Madagascar's national conservation strategy, a bureaucratic endeavor that constituted a meaningful step toward building environmental awareness across the government. The decree that legislated the strategy also established the Commission Nationale de Conservation pour le Développement (CNCD) (National Commission for Conservation and Development), assisted by a

Comité Technique Permanent (CTP) (Permanent Technical Committee) that reported to the director general of planning (Repoblikan'i Madagasikara 1984). This strategy formed the basis for a 1985 WWF-International and IUCN-sponsored conference, the idea for which Petter had first raised in Jersey in 1983.

With the increased involvement of policy makers and politicians from Madagascar and overseas, the conference in 1985, entitled the Conférence de Madagascar sur la Conservation des Ressources Naturelles au Service du Développement (usually referred to in English as the Second International Conference on Conservation and Development in Madagascar), transferred the challenge of addressing Madagascar's environmental degradation from the scientific into the political realm.[11] Many people recall this meeting as the point at which Prince Philip, then the WWF-International president, confronted President Ratsiraka by saying, "Your nation is committing environmental suicide" (Jolly 2004, 210; Kull 1996, 61).[12] The main sessions took place at the Ministry of Foreign Affairs and focused on four priority issues: soil conservation, protected areas, fisheries, and education. In contrast to the conference in 1970, the one in 1985 was perfectly timed to meet rising donor interest in Madagascar: funds did materialize to implement its recommendations, and these shaped the development of the World Bank–led Madagascar NEAP. As one attendee recalled, "That was the defining moment, at the 1985 meeting. From there, the World Bank took over and started to think about putting together these national environmental action plans."[13]

Importantly, even as the conference brought the issue of Madagascar's environment into the political realm, scientists continued to influence conservation policy. In the three years between the conference and the issuance of the action plan, a number of critical events occurred. Two preconferences, both held in the Solimotel in Antananarivo, concentrated on scientific research. The first, sponsored by the Ministère de la Recherche Scientifique et Technologique pour le Développement (MRSTD) (Ministry of Scientific Research and Technology for Development) and organized by Lala Rakotovao, the director of environmental sciences research, concentrated on the state of research on forest ecosystems in Madagascar. The second, organized by Mittermeier and Richard and sponsored by the IUCN Species Survival Commission (SSC), aimed to develop a list of Species Conservation Priorities in Madagascar. In a memo to potential participants the organizers wrote, "Special emphasis should be placed on identifying the highest priority species that are in the greatest danger of extinction, and also the most important parks and reserves. This information will be incorporated into a list of recommendations to be presented at the National Conservation Strategy Conference the following week, and will also serve as the basis for an IUCN/SSC Action Plan on Species Conservation Priorities in Madagascar" (Mittermeier and Richard 1985). Indeed the ensuing scientific assessments informed the development of a conservation action plan.

The 1985 conference and associated side meetings furthered informal collaborative relationships among scientists, donors, and Malagasy policy makers. In a 1985 memo to researchers wanting to work through the IAGS, Chairwoman Lee Durrell wrote, "I urge each of you who cannot attend the meetings to provide me with something I can present, so that, as foreigners, we can show that we are united in our aim to study, get results and therefore help sustain Madagascar's unique natural resources" (Durrell 1985). The IUCN Primate Specialist Group of the SSC was instrumental in pushing for attention to primate conservation in Madagascar, and after this conference WWF-U.S. began fund-raising.

In October 1986 WWF officials organized a trip to Madagascar for family and staff members of the W. Alton Jones Foundation, which was started in 1944 by W. Alton "Pete" Jones, an oil executive with the Cities Service Company. In addition to the foundation head, Patricia Jones Edgerton, the visitors included the executive director of the Geraldine Dodge Foundation, Scott McVay, and his wife as well as Olga Hirshhorn, the art patron and wife of the founder of the Hirshhorn Museum and Sculpture Garden, Secretary S. Dillon Ripley of the Smithsonian and his wife, Mary, and the *Washington Post* reporter Henry Mitchell, who went on to write articles about the importance of Madagascar that helped to activate USAID funding for the country's environmental issues (Mitchell 1987).[14] Mittermeier and Lovejoy escorted the group together with Jolly. The W. Alton Jones Foundation subsequently gave US$500,000 to WWF to get the conservation program in

Madagascar going in 1986–87 and to promote the development of a national conservation policy.[15] This trip led to some of the early U.S.-based seed funds for conservation and forged additional personal relationships that helped to advance the conservation agenda in Madagascar.

A subsequent series of related events focused on lemur conservation consolidated the advocacy efforts of scientists and conservation NGOs. In April 1986 the New York Zoological Society (NYZS) (now known as the Wildlife Conservation Society) hosted a meeting on St. Catherine's Island in Georgia that brought together representatives of various American and European conservation organizations to discuss the status of lemurs both in captivity and in the wild. The participants proposed a second meeting called "The Promotion of Ecology, Conservation and Development in Madagascar" that took place on St. Catherine's in May 1987 and concentrated on protected areas, captive breeding, research priorities, and training Malagasy researchers (Mittermeier 1987). Mittermeier, as the chairman of the IUCN/SSC Primate Specialist Group and director of the WWF-U.S. Primate Program, invited several Malagasy dignitaries to attend.[16] A number of zoos wanted lemurs for captive breeding, and the meeting's sponsors included WWF-U.S., the NYZS, JWPT, the San Diego Zoological Society, the Los Angeles Zoo and the Greater Los Angeles Zoo Association, the Missouri Botanical Garden, the Saint Louis Zoo, and the Duke Primate Center (Mittermeier 1987). An attendee at the meeting recalled, "[The meeting] was ostensibly about animals for zoos, but really it was about the U.S. expression of concern and the Malagasy opportunity to see the U.S. interest in zoos and animals."[17] The meeting ended with the signing of the "Convention on Collaboration with Respect to Endangered Malagasy Fauna" between the Malagasy government and various zoos, which stated that lemurs could be exported only within the context of skilled captive breeding programs and commitment to building capacity in Madagascar concurrently (Convention for Collaboration 1987).

The tensions between conservation and development goals that had arisen in 1970 were felt in these discussions as well, foreshadowing the intensifying struggle between conservation interests and Malagasy concern with economic development (see chapter 5).

Minister Randrianasolo closed the St. Catherine's meeting by underscoring the importance of integrating nature conservation and sustainable use: "Our national conservation strategy is categorical on this theme. This document expounds, in straightforward terms, that the need for sustainable development is integral to the concept of conservation" (Randrianasolo 1987).

Building Informal Relationships

After the formal meeting, a group that included the Malagasy officials, Jolly, Mittermeier, and others toured zoos in the United States, including the Duke Primate Center, the Washington National Zoo, the San Diego Zoo, the San Diego Wild Animal Park, the Los Angeles Zoo, the Huntington Botanical Gardens, the Saint Louis Zoo, and the Missouri Botanical Garden. This trip strengthened connections among various Malagasy ministries that had previously competed as well as between these ministries and foreign and Malagasy scientists. Above all, it afforded WWF needed access to Minister Randrianasolo. The final day of the trip brought a pivotal moment in Madagascar's environmental history. Mittermeier had brought along a document entitled "A Draft Action Plan for Conservation in Madagascar," which proclaimed "Madagascar the single highest major conservation priority in the world." It included a set of recommendations for the country's highest conservation priority areas and gave five-year budget estimates for each proposed project (Mittermeier 1986, 3). Its stated goal was

> to assist in the implementation of Madagascar's National Conservation Strategy by providing guidelines as to how the various international conservation organizations concerned with conservation of biological diversity can become involved in identifying the specific, highest priority projects required over the next five years. It is hoped that this document will be instrumental in finding the international support needed to carry out these essential conservation activities, that it will make a significant contribution to the survival of Madagascar's rich and unique biological diversity, and, in keeping with the spirit of the November

1985 Conference, that it will make a contribution to the long-term well-being of the Malagasy people, whose future is so closely intertwined with that of the rich natural resources that their country is so fortunate to possess. (Mittermeier 1986, 6–7)

Recalling the event, Mittermeier said, "We were driving around in a car in Los Angeles, and I finally thought that the timing would be right to present the action plan to the Minister. I turned to him and said [in French], 'Mr. Minister, I would like to officially present the Conservation Action Plan for Madagascar.' I just handed it to him. He laughed and said, 'So this is what you call an official presentation?' After much laughter by everyone in the car, he said, 'OK, I will read it tonight.'"[18] Jolly, in turn, summarizes the events that ensued on the final day of the trip. After calling the group together, "the Minister slapped down onto the bed Russ Mittermeier's Environmental Action Plan. . . . 'This,' said the Minister, 'is a Plan of Environmental Action for Madagascar. It is a five-year plan covering all proposed projects. . . . Now, can we have it sent to the inter-ministerial committee: the persons here present, who have undergone the intimacy of St. Catherine's? . . . The document will go as a draft to this committee; we will modify it, and then it can be presented as a work of collaboration between Malagasy and vazaha [foreigners] to all the others concerned. With the President's help, I think we should pass that final stage in about five minutes, a few months from now" (Jolly 2015, 94–95; see also Jolly and Sussman 2007). In this moment the previous ten years of conferences, meetings, and research agreements coalesced into a Madagascar government agenda, and the action plan subsequently informed the biodiversity portion of the Madagascar NEAP.[19]

This document was not the only outcome of the St. Catherine's trip. After a visit to Paris to meet Petter as well as IUCN and WWF-International officials, the Malagasy policy makers made a final trip to JWPT (Jolly and Sussman 2007; Mittermeier 1987). There, Rakotosamimanana negotiated a Tripartite Commission of the Ministries of Higher Education, Scientific Research, and Water and Forests to evaluate foreign research requests. It mandated that research be done cooperatively with Malagasy counterparts and that equipment "ranging from microscopes to 4 x 4 vehicles" be donated (Jolly and Sussman 2007, 32). Then, in 1988, the Madagascar Fauna Group (MFG) formed as an international consortium of twenty-one zoos and research institutes in the United States, Europe, and Great Britain that aimed to conserve Madagascar's endangered species in line with the St. Catherine's agreement. It managed the Ivoloina Zoological Park near Toamasina and the Betampona Reserve, where captive lemurs were released, as well as aided the Tsimbazaza Botanical Gardens and Zoo in Antananarivo. Originally led by the San Francisco Zoo and the Duke Primate Center, the MFG moved its headquarters to the Saint Louis Zoo in 2003, and it has remained active in conservation politics (Jolly and Sussman 2007; MFG 1994; Sargent and Anderson 2003).[20]

As these meetings among scientists and Madagascar policy makers were happening, advocates in Washington, D.C., including from WWF-U.S., were mobilizing U.S. political interest both in biodiversity and in Madagascar. In 1988 Mittermeier chaired a World Bank Task Force on Biodiversity that raised awareness of the issue within the World Bank. That same year the Smithsonian Institution signed a Memorandum of Understanding with the Madagascar scientific research ministry, with the goals of promoting bilateral cooperation, facilitating research permits, and promoting research exchanges in various scientific fields (MRSTD and Smithsonian Institution 1988). Beginning in April 1989 the Smithsonian assembled a group of American scientists as well as policy makers from the Smithsonian, the World Bank, and USAID to discuss strategies for protecting Madagascar's biodiversity and for moving forward the NEAP (Smithsonian Institution 1990).[21] Many of the American scientists working in or interested in Madagascar at the time participated in these meetings, and, as one scientist recalled, "We—the research professors, the policy makers, decision makers and the finance people—had meetings called together by the Smithsonian Institution. We all contributed to a certain extent to [the Madagascar] environmental action plan by going to these meetings, by discussing these things, and sometimes writing too. We didn't know it then, but the NEAP was the result of all these meetings."[22] Like the trips and events discussed above, these meetings solidified informal interpersonal relationships and institutionalized initial policies, laying the foundation not just for the conduct of scientific research in Madagascar but also for the influence of U.S.-based scientists and conservation NGOs in Madagascar environmental politics.

The Perfect Model

By the time the World Bank adopted environmentalism in the late 1980s these science-policy-advocacy networks were well established, and the bank provided the finances to transform this growing scientific concern into a political reality. In 1987 the World Bank president Barber Conable announced, in an address to the World Resources Institute (WRI), that the bank would begin paying more attention to environmental issues, specifically by creating an environment department, undertaking countrywide national environmental assessments, and funding environmental programs (Conable 1987). The World Bank began by producing internal Environmental Issues Briefs and Country Environmental Strategy Papers (Falloux and Talbot 1993; Wade 1997). The environmental officers who led the NEAP process, François Falloux and Lee Talbot (1993), credit USAID for developing the model for the bank's assessments: "The first such descriptions in Africa were developed in the 1970s by USAID. Initially these efforts were generally brief desk studies, which summarized the environmental conditions of the country to provide a basis for USAID planning. Later, in the form of Country Environmental Profiles they have sought to assess a country's natural resource potential in relation to economic growth and development. The profiles have helped to establish an information base that can be used in planning and policy development" (13).

In 1992 the World Bank began requiring all borrowing countries to produce NEAPs in order to qualify for structural adjustment lending (Goldman 2005; Marcussen 2003). By 1996 more than ninety countries had started a NEAP process, and seventy-four plans had been completed (World Bank 1996). Like the earlier environmental assessments, NEAPs were supposed to identify environmental problems, analyze their underlying causes, and recommend actions to address them, the goal being to mainstream the environment into the overall development planning process of a country (Greve, Lampietti, and Falloux 1995). They were also intended to provide a mechanism with which to coordinate donors as well as scientific organizations, NGOs, and other institutions around complementary and integrated actions.

Reflecting the emphasis on involving civil society in policy processes that characterized the modified neoliberalism of the late 1980s (Hart 2001; Mohan and Stokke 2000), the World Bank pushed the Malagasy government to involve private and nonprofit organizations in the development and implementation of its NEAP (Froger and Andriamahefazafy 2003; Sarrasin 2007b). The bank emphasized decentralized awareness building among both populations and government authorities in order to reinforce "country ownership" and to involve "the population" (Andriamahefazafy and Méral 2004; Falloux and Talbot 1993; Froger and Andriamahefazafy 2003). To this end the plans were to be "holistic," "processual," "country owned and driven" (instead of donor-driven), and "participatory": "A 'process' more than a 'product,' a NEAP seeks to provide a framework for integrating environmental considerations within the overall economic and social development of a country. As a truly national enterprise this process should be taken over and orchestrated by each interested country; it is not done for the country by a donor. The government and the civil society are partners and wide public participation is essential" (Falloux and Talbot 1993, 1).

Because Madagascar already had the National Conservation Strategy of 1984 and governmental mechanisms to coordinate its implementation in addition to well-established relationships between scientists and policy makers, the country was an ideal place to showcase the World Bank's new environmental agenda. Madagascar afforded the bank an opportunity to appease the influential U.S.-based environmental groups who were concerned with biodiversity loss and deforestation. At the same time, the environmental agenda offered the Madagascar government an avenue to attract much-needed foreign exchange in the context of IMF restructuring (Horning 2008a; Sarrasin 2005). As a result, "ten years after 1975 there was basically a 180-degree turn, and in 1985 the government realized that environment was the fishhook for foreign aid and decided to do what the donors wanted in order to get foreign aid."[23] As it turned toward the Anglophone Western world, the Madagascar government adopted not just neoliberal policies but also the international environmental agenda. The Madagascar NEAP became the nexus for the negotiation of diverse agendas, spanning the World Bank, USAID, international conservation NGOs, scientists, various Madagascar government agencies, and others.

Other countries began using the Madagascar program as a model for coordinating donors and government around a unified environmental agenda and as a test for the international donor community's capacity to protect the global commons. Much of the overarching NEAP design was developed on the basis of the Madagascar plan (Mercier 2006). A former senior Malagasy official recalled that as a fifteen-year, donor-government coordinated plan it represented an entirely new way of providing foreign aid, and as a result, "a lot of countries and international organizations were interested in the [Madagascar] Environmental Action Plan."[24] In this regard its designers saw an opportunity to shape not just Madagascar's future but also that of the world. The foreword to the 1988 draft NEAP states, "The case of Madagascar presents the international community an opportunity to create and implement an original solution for development assistance that will preserve this biological diversity—a diversity which is part of the common heritage of all humanity. If successful, such a solution will serve as a future model for other countries" (World Bank, USAID, et al. 1988, 2).

Following the conference in 1985 and drawing on the 1984 strategy, the Madagascar government created an interministerial committee and a small planning unit to implement the strategy. The temporary Cellule d'Appui au Plan d'Action Environnementale (CAPAE) (Support Center for the Environmental Action Plan), staffed primarily by private consultants, coordinated its preparation. Roughly 150 Madagascar government analysts, academics, and consultants and some 40 international environmental experts were involved in its development (Brinkerhoff and Yeager 1993; Sarrasin 2006a).

World Bank missions in 1987 and 1988 under the guidance of Falloux then pushed the NEAP forward (Brinkerhoff and Yeager 1993). The first World Bank NEAP planning mission was in October 1987; topical working groups started in late 1987 to map out priorities; and a final mission in March 1988 brought together representatives of USAID, WWF-International, the World Bank, and United Nations Educational, Scientific, and Cultural Organization (UNESCO) as well as French and Swiss consultants (Sarrasin 2001, 2007b; World Bank, USAID, et al. 1988). The working groups presented final recommendations at a conference in Paris in May 1988, and the draft NEAP was published in July 1988. Through these missions and working groups the World Bank enlisted bilateral donors, NGOs, scientists, and others in its vision for the NEAP; the NEAP became an "obligatory passage" for engaging in Madagascar environmental politics (Sarrasin 2007b).

Engaging "Civil Society" and Avoiding the State

Foreign interests heavily influenced the negotiations. The consultant Jean-Roger Mercier (2006) recounts efforts to ensure that foreigners did not dominate the process: "While the original team was essentially composed of international experts, we rapidly co-opted several Malagasy experts and anchored our contacts with the Malagasy Government, which was involved at the highest level, [with] the then Prime Minister Victor Ramahatra bringing an incredibly pertinent vision to this NEAP preparation. Cooperation with the international NGOs was a given, with WWF having a particularly strong and competent involvement from the onset" (50). Brinkerhoff and Yeager (1993) expand on the influence of international conservation groups:

> The majority of support for environmental efforts came from international nongovernmental groups such as WWF, and U.S. universities including Duke University, North Carolina State University, Yale University, Washington University, and the Missouri Botanical Gardens.... In the early stages of the [Madagascar] environmental movement, it appeared that Malagasy government officials, scientists, and development agents would play a lead role in orchestrating the effort. Over the long run, however, the international conservation groups and donors became key players

in promoting and encouraging continued action, working with a core group of Malagasy environmentalists. Due in large part to the resources they were able to commit to the effort, international donor agencies continued to play a major role as the Government of Madagascar embarked on the environmental planning process. (7)

Building on the strong networks and advocacy around biodiversity that had already been established, conservation organizations were able to take advantage of this moment to channel political and financial support to biodiversity. Because WWF had the background information, the World Bank asked it to write the biodiversity portion of the Environmental Action Plan.[25] The resulting conservation agenda drew on the conservation strategy of 1984 (Repoblika Demokratika Malagasy 1984), Mittermeier's Conservation Action Plan (Mittermeier 1986), and the biological surveys conducted by WWF-International (Nicoll 1988).

The massive mobilization of nongovernmental personnel allowed the World Bank to tout the program as participatory (Bhatnagar and Williams 1992), even though the participants were primarily foreign- or Antananarivo-based. In their review of the process, Falloux and Talbot lament that few Malagasy NGOs were involved, the notable exception being the Federation of Malagasy Churches, which ultimately "played a key role in disseminating information and mobilizing support for the NEAP." They continue, "A major problem at the start was that the NEAP development was confined to the intellectual and technological circles in the capital, Antananarivo. To remedy this, albeit almost too late, a series of regional seminars were organized" (Falloux and Talbot 1993, 102–3). However, just as happened with SAPM decades later, pressure to speed up the process hindered regional consultations: "In an effort to maintain the momentum of the analysis, input from politicians, government officials, and farmers outside of the capital was not solicited" (Brinkerhoff and Yeager 1993, 9). This Antananarivo-centric approach both reflected and reinforced a long history of struggles over land between Merina and coastal people—an issue that resurfaced throughout the NEAP and in the twenty-first century expansion of protected areas (fig. 3.3).

Figure 3.3. Antananarivo. Photograph by Desmond Fitz-Gibbon.

Donors sought to simultaneously avoid and engage the state. Although the CAPAE was sponsored by the Directorate of Planning and reported to the CTP (which worked under the authority of the CNCD), it was financially and administratively autonomous from the government (Falloux and Talbot 1993; Pollini 2011). Creating it as a parastatal organization allowed the World Bank to pay higher salaries than the government, where structural adjustment was holding down civil service salaries (Jolly 2015) and "to maintain the balance of power in favor of [donor] 'experts' while facilitating incentives for the government and Malagasy public administration in the project" (Sarrasin 2007b, 442, my translation). The organization depended on foreign donors, including the World Bank, USAID, the Swiss aid agency, United Nations Development Program (UNDP), UNESCO, and WWF-International, for all of its finances, including salaries (Pollini 2011; Sarrasin 2007b).

In another failed effort to promote Malagasy control, the donors committed to locating a Multi-Donor Secretariat (MDS) to coordinate the eleven donor agencies that would finance the first phase of the Madagascar NEAP in Antananarivo (Cooperation Suisse et al. 1989; World Bank 1989). However, in 1989 USAID agreed to finance an MDS at the World Bank in Washington.[26] The justification for moving the MDS to Washington was to facilitate coordination with donor and NGO headquarters outside of Madagascar and to let the newly created Office National pour l'Environnement (ONE) (National Environment Office) coordinate those within Madagascar. In fact, the MDS eventually became a conduit of NEAP information and experience among countries around the globe, and it helped to coordinate NEAPs across a number of African countries. In the second phase of the NEAP, a donor-financed, Madagascar-based Secrétariat Multi Bailleur (SMB) (Multi-Donor Secretariat) replaced the MDS and served as an interface between donors and the Madagascar government on environmental funding and priority setting (Brinkerhoff and Yeager 1993; Brinkerhoff 1996; Falloux and Talbot 1993; Greve, Lampietti, and Falloux 1995; Lindemann 2004; World Bank 2003c).

By creating new institutions outside of the government the donors could control the priorities and pace of the program, but they also had to sell their agendas to key officials in the Madagascar government. Even as Falloux and Talbot (1993) commend the "wisdom not to entrust the NEAP preparations directly to the existing governmental structure" (36), they admit that the lack of parliamentary involvement resulted in a slowdown of the NEAP's formal acceptance. The CAPAE tackled the challenge of selling the NEAP to the government in two primary ways. First, advocates hired media consultants to educate the population about environmental issues and to build popular support for the NEAP. The media attention helped to enlist government officials (Sarrasin 2007b). As Falloux and Talbot (1993) recall, "The president was obligated to enter the arena after the showing of an excellent series of televised environmental episodes produced by Radio Television Malgache. The environmental series was intended to strengthen public opinion in favor of the NEAP. This coincided with the start of the president's re-election campaign, and under the journalists' pressure the president was obliged to present his environmental polices to the public" (34–35).

Second, they made a broad effort to articulate the economic importance of the environment. Experts in different sectors were asked to estimate the "average loss incurred when one acre of natural forest was robbed of its natural function and valuable contents" or "the loss of productivity resulting from soil erosion" or the "annual loss of irrigated land surface because of desertification" (fig. 3.4). Once they did this, "only the costs which were easily assessed and quantifiable were retained." The result was a rough (and distorted) estimated cost of environmental degradation at US$150–300 million, or 5–15 percent of the gross national product. This number was sufficient to enlist the prime minister's endorsement: "On the basis of the alarming estimates of the costs of the environmental degradation the Prime Minister joined the director of planning as a sponsor of NEAP" (Falloux and Talbot 1993, 34), and "the Minister of Economics and Planning ... became the chief 'supporter' of the NEAP" (Falloux and Talbot 1993, 42–43). Thus the ability to quantify both the costs of environmental degradation and the benefits of environmental conservation was instrumental in enlisting government agencies. Twenty years later this same ability

Figure 3.4. Lavakas *from soil erosion. Photograph by Catherine Corson.*

to quantify the costs and benefits of forest loss would convince President Ravalomanana to triple Madagascar's protected areas.

Ironically, in this model, which embodied the colonial legacy that promoted commercial extraction while condemning tavy, conversion of land (for timber) was considered a positive land use while shifting cultivation was not. Initial attempts to quantify the costs of continued environmental degradation attributed 80 percent of the costs to shifting cultivation, calculated as the opportunity cost of not being able to use forest land for timber harvesting (Larson 1994). Thus a bias against land use for subsistence and for commercial export was introduced into conservation planning from the beginning, and it is unsurprising that one of the main goals of the Madagascar NEAP was to discourage shifting cultivation, sedentarize farmers, and encourage them to invest in soil conservation, agroforestry, and reforestation via integrated development in the zones surrounding protected areas.

Debating the Balance: Lemurs Versus People

The NEAP had the potential to set the stage for a broad environmental program for Madagascar, but the operationalization of the plan depended almost entirely on foreign aid donor funding. The NEAP preparations alone cost US$1.3 million, 40 percent of which came from the World Bank, 26 percent from USAID, 17 percent from the Madagascar government, 11 percent from the Swiss government, and 6 percent from UNDP (Greve, Lampietti, and Falloux 1995). Mercier (2006) recalls the early implementation process: "Our first order of business was to define the NEAP's scope. We cast the net very widely and did not limit ourselves to conservation, though conservation was both the reason why Madagascar was so famous and courted internationally and the biggest motivation behind the preparation of the NEAP" (50). The final NEAP of July 1988 laid out the following priorities: (1) soil conservation; (2) protection and management of biological diversity; (3) development of ecological tourism; (4) establishment of an institutional framework for the environment; (5) urban environment; (6) mechanisms for the management, protection, and continuous monitoring of the environment; (7) development of environmental research; and (8) development of education, training, information, and sensitization of the protection and management of the environment (World Bank, USAID, et al. 1988). While issues such as biodiversity conservation, urban environment, and soil conservation continued to be emphasized, priorities in early planning meetings such as human health, marine issues, and energy were marginalized in favor of education, research, monitoring, and tourism.[27] Overriding these conflicts among donors was an ongoing clash between conservation and development goals: as a government official recounted, "We were concerned with development, but the donors were interested in conservation."[28] As the holders of the purse strings, the foreign aid donors quickly began reshaping the plan's priorities.

Once the NEAP was accepted in 1988, subsequent multidonor missions in 1989 negotiated its implementation. Again, foreign and nonstate actors dominated. The World Bank meeting in Madagascar in July 1989 was a pivotal moment, when Swiss, American, Norwegian, and German donors on the multidonor mission released a joint memo to the World Bank that critiqued the priorities for the first phase of the program laid out in the World Bank's summary of the 1989 Donor Evaluation Mission (World Bank 1989). The confrontation took place at the end of the first week of the joint donor mission, after the participants at the mission had spent a weekend at the Périnet reserve. Jolly (2015) describes what happened: "The aid donors sat in a grim clump at the far end from François [Falloux]. When François called on Hans Hürni [with the Institute of Geography at Berne University], Hans just rose with a paper from the donors in his hand, walked silently the length of the table, put it down in front of François and walked silently back" (111; see also Jolly 2004, 115).

Challenging the World Bank's proposed plan of action, the memo argued that the proposed program had lost focus, that biodiversity should remain the central component, and that a number of the proposed components of the project were too ambitious to be accomplished: "Due to insufficient institutional capacity and technical experience, the MDG [multidonor group] suggests scaling down the soil conservation, teledetection/cadastre and education components.... In addition, the MDG strongly feels that inadequate training and institutional capacity is the single most significant constraint to improved environmental management in Madagascar, and needs to be addressed in a more coherent way within each project component." It underscored that the biodiversity section "continues to be the most coherent component of the project, and should serve as a focal point for other project activities." It added that strong support for the Department of Water and Forests was essential, that the potential for tourism was overstated, that the most promising venue for recurrent cost funding was an endowment for US$5–10 million, and that land titling, education, and rural community projects should focus to a great extent on communities around protected areas. Finally, it emphasized that the proposed MDS should be based not in Washington but in Madagascar (Cooperation Suisse et al. 1989, 2–3). At this meeting also USAID announced its intention to focus on biodiversity.[29]

Again, tensions between donor interest in conservation and Malagasy interest in development surfaced: "By the next day the Malagasy counter-attacked. Viviane Ralimanga (the head of the CAPAE) herself wrote an impassioned letter saying if we thought we could just emphasize fauna and flora, we were sadly misjudging the temper of the Malagasy, as well as their needs. Philippe Rajobelina, the Deputy Director General of Planning, wrote to say that even within the biodiversity sector it was unacceptable to have more money allotted to the reserves than to peripheral development because 'There are more important primates in Madagascar than lemurs'" (Jolly 2015, 112–13; see also Jolly 2004, 215).

The negotiated outcome was a concentration on five priorities, which reflected those of the donors as a collective body: education, biodiversity conservation, soil conservation, land titling, and institutional development (Democratic Republic of Madagascar 1990, 41). The World Bank projected the costs for the first phase of the NEAP at US$85.5 million, to which donors pledged funding at a roundtable meeting held in Paris in February 1990. The figure contained an estimated US$27.7 million for biodiversity, US$12.7 million for cartography, US$9 million for land titling, US$5.1 million for research, US$12.6 million for soil conservation, and US$18.4 million for institutional support to ONE. Ultimately, donors funded 80 percent of the projects' costs, while the government of Madagascar was responsible for the remaining 20 percent (Horning 2008a, b).

Nevertheless, by 1991 the programs proposed or in place for biodiversity totaled over US$60 million, more than US$50 million of which came from USAID, with UNDP, German, Norwegian, and proposed British aid making up the balance (Greve 1991b). Moreover, programs like mapping, land tenure, research, and information were often oriented toward biodiversity and forests programs (Hufty and Muttenzer 2002). Second, even within the biodiversity program there remained tension between how much to focus on conservation and how much to integrate development, and while ICDPs offered a balance (see chapter 6), by the second and third phases of the NEAP they had given way to large-scale biodiversity prioritization and landscape planning. The first MDS newsletter states, "One of the major issues facing EP1 as it approaches implementation is how to achieve the correct balance between biodiversity and natural resource conservation, scientific research and development activities for the buffer zone populations" (Greve 1990, n.p.). At the annual meeting of the Steering and Monitoring Committee in 1990, the donors decided that the biodiversity conservation program needed a new name, one that would "reflect the need for programs developed under this component to ensure a balance between the development of human communities and the conservation of threatened biodiversity" (Greve 1991b, n.p.). Yet even as donors agreed that it was important to integrate development with conservation, a long-standing emphasis within the biodiversity program on biological inventories, identification of conservation priorities, and the expansion of protected areas remained. The program highlighted the following priorities: (1) the inventory of Malagasy ecosystems and identification of conservation priorities; (2) the establishment of a network of fifty parks and natural reserves and improved forest protection in targeted areas; (3) the involvement of the surrounding populations in park management and the creation of economic or social benefits for populations around parks; (4) the mobilization of international financial and technical support; (5) the establishment of a national institutional framework through the creation of Association Nationale pour la Gestion des Aires Protégées (ANGAP) (National Association for the Management of Protected Areas);[30] and (6) the development of ecological tourism with the aim of covering the recurrent costs of parks through entry and concession fees (Falloux and Talbot 1993). These priorities remained core emphases in the conservation program for decades, as did the tension between conservation and development interests (see chapter 5).

Science, Power, and Governance

The history of Madagascar's conservation politics highlights the value of attending not just to official donor coordination structures, conferences, and legislation but also to the informal interactions among the individuals engaged in them. From the mid-1970s through the launch of the NEAP in the late 1980s an assemblage of dedicated scientists, NGOs, donors, and bureaucrats worked together in both informal and formal ways to facilitate scientific research and promote conservation in Madagascar. Through meetings, conferences, trips, letters, agreements, and action plans, they circulated ideas, crafted narratives, and developed policies that laid the foundation for Madagascar's subsequent environmental program. At particular moments—some of which were unplanned—ranging from major conferences to trips to park lodges they shifted the political playing field in critical ways. Their explanations for Madagascar's environmental crisis as well as for the best solutions to it became institutionalized, not only via the official policies that the World Bank, the Madagascar government, foreign donors, conservation NGOs, and others crafted but also through the personal relationships they developed during this period, relationships that continue to influence environmental politics in Madagascar to this day. Throughout this process, although numerous actors advocated for integrated conservation and development approaches, the political, scientific, and financial strength of the conservation lobby often overrode the push for more comprehensive or integrated conservation and development approaches.

The program's concentration on biodiversity reflected not just the efforts of a group of individuals and the timing of the World Bank's environmental interest but also the particular relations of governance brought about by the rise of neoliberalism. The neoliberal reduction of the state, the participatory turn in international development, the World Bank's adoption of the environment as a central issue, and the rising global attention to biodiversity enabled this assemblage to transform Madagascar's conservation agenda from a scientific issue to a political one. The push for participatory policy development legitimized nonstate actors' influence on the environmental priorities even as the participation was primarily by Antananarivo-based and foreign actors. Likewise, the reduction of the Madagascar state under structural adjustment and the resulting lack of state capacity and accompanying need for foreign exchange created the conditions under which the Madagascar government had to embrace donors' and NGOs' priorities. In this sense the rise of biodiversity conservation was intimately intertwined with that of neoliberalism.

CHAPTER FOUR

Tracing the Roots of Neoliberal Conservation

The beauty of earmarks is . . . you just need three or four
people to create a [US]$300 million earmark.

—INTERVIEW WITH A FORMER USAID OFFICIAL, July 7, 2006

It is easier to do biodiversity overseas than in this country
because the conflicts don't involve constituencies of
Congress.

—INTERVIEW WITH A FORMER USAID OFFICIAL, August 3, 2005.

SINCE THE 1970S AN evolving group of environmental NGOs has
urged the U.S. Congress to fund environmental foreign assistance.
Dynamic power relations between this group and USAID have led
to ideological changes in what has constituted "the environment" in
the agency's programs. At its height, USAID's environmental program encompassed a range of issues. However, as the assemblage
of NGOs pushing environmental foreign aid shifted from a loose

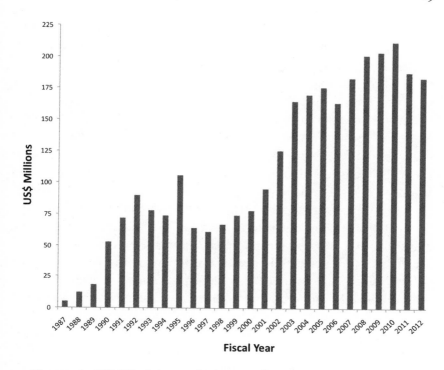

Figure 4.1. *USAID Assistance for Biodiversity Conservation, 1987–2012. Source: USAID Biodiversity Conservation and Forestry Programs Annual Reports.*

alliance of environmental advocates in the 1970s to a coalition of biodiversity conservation NGOs in the twenty-first century, so too did the agency's priorities.[1] Building on the previous chapter's analysis of how a scientifically grounded movement became a successful political campaign, in this chapter I trace the history of USAID environmental funding, focusing on the expansion of its biodiversity conservation program, which grew from US$5 million in fiscal year (FY) 1987 to an estimated US$184 million in FY 2012 (fig. 4.1) (USAID 1994c, 2009, 2013).[2] I argue that the rising prominence of biodiversity with USAID's environmental portfolio was made possible by the neoliberal reduction of the state, a development that opened up room for nonprofit and private actors to influence state policy. As USAID turned to nonprofit and private actors for assistance in policy making, program design and implementation, evaluation, and enforcement, these actors also became key advocates for USAID's environmental program on Capitol Hill, and the agency's environmental program narrowed to mirror the priorities of its political champions.

Four critical historical conjunctures, loosely defined by decade, underpinned this transformation. First, in the mid-1970s environmental advocacy groups sued USAID to compel the agency to address the environmental impacts of its activities and then lobbied the U.S. Congress to require it to fund environmental projects. USAID's initial environmental program began as a response to this lawsuit, and it focused on addressing negative environmental impacts and managing environmental resources for rural livelihoods and economic growth. Second, in the context of increasing data on forest loss and associated species extinction in the early to mid-1980s NGOs successfully triggered congressional action that required USAID to fund biological diversity. Concurrently, Congress, in reaction to the Reagan administration's push for private sector–led development, began encouraging USAID to fund NGOs to implement the resulting programs. In the late 1980s, as many of the advocacy groups that had initially catalyzed USAID's environmental agenda shifted their focus to the multilateral development banks, the conservation NGOs funded through USAID's new biodiversity initiative emerged as the primary political advocates of the agency's environmental programs.

By the mid-1990s, the third historical conjuncture, USAID's biodiversity program had grown significantly, buttressed by the rise of the sustainable development discourse and the Clinton administration's use of foreign aid to address global environmental issues in the post–Cold War era. The Clinton administration and the Republican Congress implemented reforms to reduce the size of the government, increasingly turning over funds and project management to private and nonprofit organizations, which fueled the growth of the conservation NGOs. Prompted by further budget reductions to USAID, these NGOs organized to protect biodiversity funding. As they grew substantially in the 1990s they capitalized on idealized visions of themselves as representatives of civil society operating to counter the private sector, even as they began nurturing corporate partnerships.

In the early twenty-first century, the fourth historical conjuncture, they began attracting corporate sponsorship and bipartisan political endorsements, which led to substantial increases in USAID's biodiversity funds. They did so by defining the environment as biodiversity over there, to be protected in parks, away from competing economic and political interests and in foreign countries. This environmental framing elided controversial issues and created an avenue by which U.S. politicians and corporate leaders could become "environmentally friendly" without angering their constituents, funders, or political allies.

I begin by chronicling how the broader political context of U.S. foreign aid shaped both the strategies that environmentalists used to advocate for their interests and how USAID responded. I then trace the transformation of the NGO environmental foreign aid lobby and associated rise of biodiversity in USAID's environmental agenda. I conclude with a discussion of how biodiversity's ascendance impeded the agency's ability to address drivers of environmental degradation and ultimately created new spaces for capital accumulation.

The Context for Environmental Foreign Aid

The broader political context of U.S. foreign aid shaped both the strategies that environmental advocates used to advance their interests and how USAID responded. Four aspects merit specific mention: USAID's marginalization within the executive branch; the rising role of Congress in shaping foreign aid priorities; the lack of public support for international development; and the introduction in the 1970s of the basic needs era.

U.S. foreign assistance programs came into being after World War II as a crucial instrument of the emerging U.S. economic and political empire. They concentrated on economic and military aid to Europe until in 1961 the FAA refocused foreign aid on the developing world and pulled together previously scattered activities into one agency, USAID. However, as a subcabinet agency, USAID is still often sidelined within the federal government, even with respect to development policy. It is responsible for only a small percentage of foreign aid: the Treasury Department, the U.S. Department of

Agriculture, the Department of State, the Department of Defense, and, more recently, the Millennium Challenge Corporation all administer their own aid programs.

During the Kennedy and Johnson administrations the White House built congressional backing for foreign aid around the idea that it could help contain communism, promote democracy and education, and foster trade for American companies. However, as public protest against the Vietnam War grew, members of Congress increasingly challenged the president's foreign aid priorities, and in 1971 Congress rejected the USAID FY 1972 and 1973 authorization bills. A number of scholars have argued that after the Kennedy administration there was no comprehensive congressional base of support for U.S. foreign aid (e.g., Berg and Gordon 1989; Hoben 1989; Porter 1990). Concomitantly, there has been limited public approval of foreign aid. Studies in the mid-1990s reported that 65–75 percent of Americans thought that the United States spent too much on foreign aid. While more recent studies have reported a smaller number, 40–47 percent, they still indicated that a sizable number of Americans preferred to spend less on foreign aid (Kull 1995; Lancaster 1999; Milner and Tingley 2013; USAID 2002a).

The lack of public and congressional backing has meant that USAID depends on the efforts of special interest groups, including American farmers (who have traditionally endorsed the Food for Peace program), humanitarian groups, universities, private sector companies, and conservation groups to mobilize congressional support for its funding. In 2004 alone 143 companies and organizations lobbied on USAID programs (Center for Public Integrity 2008). NGOs, including arms of major churches, universities, service delivery groups, think tanks, and advocacy groups, have been staunch champions of foreign aid. They have all come together under the auspices of NGO umbrella groups like InterAction to advocate for foreign aid in general, but, like other lobbyists, they have often promoted specific issues and projects, contributing to the proliferation of legislative directives in the annual congressional appropriations bills, or earmarks, which require that an agency spend money on a specific country, organization, or issue. A Congressional Research Service (CRS) report of 2006 concluded that 74.6 percent of that year's foreign aid budget, or US$16 billion, was earmarked. While neither earmarks nor the increase in earmarks have been unique to the foreign aid bill, for many years it was the most highly earmarked of all the appropriations bills, the others being characterized by earmark rates ranging from 1 to 20 percent of their overall appropriations (CRS 2006).[3] Through earmarks in bills and similar directives contained in the accompanying congressional reports Congress has had a major influence on USAID's priorities and programs, and in this chapter I concentrate specifically on the politics of the biodiversity earmark.

The convergence in the early 1970s of growing public interest in environmental issues with the Basic Needs Era in international development laid the foundation for USAID's environmental program. The Basic Needs Era, that is, the decision to focus aid on the basic needs of the poor as opposed to large capital projects, is often associated with the announcement in 1973 by Robert McNamara, the president of the World Bank, of his decision to tackle rural poverty. In the United States it manifested in the New Directions legislation of 1973, which amended the FAA to require USAID to prioritize agriculture, health, education, and population planning. Although this legislation did not focus specifically on the environment, it provided a framework within which to respond to the budding American environmental movement, which gained momentum in the 1960s and 1970s, spurred in part by the publication of Rachel Carson's *Silent Spring* in 1962 and encapsulated in the celebration of Earth Day in 1970. Responding to this interest with a Keynesian emphasis on the role of the state to protect human welfare, the Democratic Congress legislated a number of environmental regulatory reforms in the late sixties and early seventies, including the National Environmental Policy Act (NEPA), which set environmental standards for government-funded activities.

In the early 1970s, media coverage of the Sahelian drought, oil shocks, and the energy crisis increased Americans' awareness of foreign environmental issues and the problems of unfettered growth. In 1972 the United Nations Conference on the Human Environment in Stockholm met to discuss industrial pollution in the developed world. That same year, expressing concerns that surfaced again during the oil crisis of 1973, the Club of Rome published its study *Limits to Growth*, which predicted that economic growth could not continue indefinitely due to the limited supply of natural resources. In 1974 Lester Brown founded World Watch Institute, and Erik Eckholm wrote the first World Watch paper, called "The Other Energy Crisis: Firewood," which documented fuelwood issues around the world (Eckholm 1975). Eckholm went on to write a book entitled *Losing Ground: Environmental Stress and World Food Prospects* (Eckholm 1976), which discussed how environmental stresses would hinder people's ability to feed themselves in developing countries. An environmental activist who helped to start the USAID program recounted, "That paper just blew people's minds, because it was at a period we were dealing with an energy crisis, 1974/75, in the United States, and all of the sudden he said there's another energy crisis out there with firewood. So *Losing Ground* was the beginning of popularizing literature on a series of environmental problems in developing countries, which included deforestation, desertification, salinization, all of which were having a direct impact on the capacity of poor people to sustain themselves and improve their economic conditions."[4] A small group of individuals introduced the arguments of this literature to Congress.

An Initial Environmental Agenda

USAID's official environmental program began as a NEPA lawsuit settlement.[5] In 1974 a number of environmental advocacy groups led by the Environmental Defense Fund (EDF) and the Natural Resources Defense Council (NRDC) sued USAID under NEPA (*EDF v. AID* 6 ELR 20121 [D.D.C. 1975]) in response to the death of five workers in Pakistan from the pesticide Malathion, provided in a USAID malaria control project.[6] The lawsuit alleged that USAID failed to undertake NEPA-required environmental impact statements and specifically targeted the agency's use of pesticides in its projects (Sirica 1975). While USAID argued that NEPA did not apply to extraterritorial activities, it agreed as part of an out-of-court settlement to establish environmental impact procedures for all of its projects, and it issued these regulations in 1976 (22 CFR part 216; 41 Fed Reg. 26913; Burpee, Harrigan, and Remington 2000). USAID's initial environmental efforts thus focused on environmental impact statements.

In 1976 U.S. tax laws changed, expanding the degree to which nonprofit organizations could lobby Congress without jeopardizing their tax exempt status (Smucker 1999). NRDC began moving away from its strategy of court action to the relatively more rapid avenue for change offered by Congress. As an NRDC advocate recalled, "We realized fairly early on that you could win a lawsuit in the environmental field, but it wasn't going to do you much good if people could outflank you in the legislature. So I think we started lobbying as more of a defensive thing, but it quickly became obvious that you could use the lobbying process to do a multipronged attack on environmental issues."[7] NRDC, together with other NGOs, including WWF-U.S., New Directions, Rare Animal Relief Effort, the National Audubon Society, TNC, and the Sierra Club, urged Congress to amend the FAA to require that USAID undertake projects specifically targeted at environmental concerns as opposed to simply avoiding or addressing projects' negative environmental impacts. Drawing on *Losing Ground*, they highlighted four key issues: deforestation, desert encroachment, threatened irrigation systems, and degenerating mountain environments (Campbell 1977; Eckholm 1976; Scherr 1978). A member of the group recollected, "I can remember vividly going up to Capitol Hill. . . . We had copies of some of the World Watch papers, and we would go to congressional staff and say, 'There are environmental problems in the developing world and we really need to address these.' And so we got section 118 [an FAA amendment] passed, and that was the beginning."[8]

The reports and letters that informed the 1970s congressional amendments emphasized the need to protect the natural resources upon which poor people in developing countries relied (Blake et al. 1980; Scherr 1978). They catalyzed the agency's environmental agenda with a focus on managing supplies of environmental goods to ensure rural livelihoods. The responding amendments to the FAA of 1977 authorized USAID "to furnish assistance for developing and strengthening the capacity of less developed countries to protect and manage their environmental and natural resources." They stated further that "special efforts shall be made to maintain and where possible restore the land, vegetation, water, wildlife and other resources upon which depend economic growth and human

well being, especially that of the poor" (U.S. Congress 1977; see
also USAID 1988c). With urging from NRDC staff and others,
Congress further amended the FAA in 1978 and 1981, requiring
USAID to undertake studies to identify major environmental
problems in developing countries and specifically highlighting the
importance of forest and soil conservation (U.S. Congress 1978,
1981a). The first USAID environmental report to Congress, in
1979, highlighted "the continuing loss of tropical forest cover, the
exhaustion of croplands, the depletion of fisheries, the advance of
desert frontiers, the adverse consequences of indiscriminate pesticide use [and] epidemic levels of environmentally related diseases"
(USAID 1979).

With this legislation in place, USAID had to develop the institutional capacity to implement environmental programs in its missions. According to one former USAID official, "Environmental
programs were run from Washington. They weren't initiated in the
field because there weren't any environmental officers."[9] To address this shortage the agency began hiring environmental staff in
the late seventies and early eighties, turning in particular to the
Peace Corps and the American Association for the Advancement of
Science program. Steen (1991) cites the impact on USAID programs made by the thousands of Peace Corps volunteers trained in
forestry and natural resource management. These volunteers often
worked on USAID or U.S. Forest Service projects in developing
countries or joined USAID upon completion of their Peace Corps
training. Early environmental officers tried to gradually integrate
the environment as an agency priority, but many longtime employees viewed it as an annoying impediment to ongoing agency operations (Hoben 1997). "Environment was always an uphill battle," a
former USAID environmental officer recounted. "We used the do
no harm angle with environmental impact statements. This was
our ace in the hole. But to sell environment, we had to make connections between rural livelihoods and environment, for example."[10] Thus USAID's early programs emphasized intervention to
protect natural resource supplies for the poor and focused on soil
conservation, alternative energy, agroforestry, and reforestation.
However, this emphasis gradually shifted toward biodiversity and
tropical forests in the 1980s.

Tropical Forests, Sustainable Development, and Biodiversity

In the late 1970s and early 1980s the campaign to save tropical forests began gaining momentum, particularly as a result of widely publicized satellite images of tropical forest loss. While USAID had funded industrial timber exploitation in the 1950s and 1960s, by the seventies and eighties it had begun investing in fuelwood conservation and antideforestation programs. A USAID-funded NRDC report in 1978 recommended that USAID build programs in four main areas: sustainable forest management, soil conservation, pesticide management, and energy (Blake et al. 1980). That same year the U.S. Strategy Conference on Tropical Deforestation reported that "the world is being confronted by an extremely serious problem . . . as the result of the accelerating loss of forest and vegetative cover in the humid and semiarid lands within or near the tropical latitudes" (U.S. State Department 1980, executive summary). A former congressional aide recalled the interest Congress showed in saving tropical forests at the time: "Every day we woke up and realized that many more football fields of the Amazon were gone and before our lives were over, it was going to be gone, and the planet was going to be completely different as a consequence."[11] One can see already in these early programs the framing of global environmental issues as a foreign problem.

In 1980 USAID launched its Forest Resources Management project, which contracted with the U.S. Forest Service to provide a variety of technical services in forestry to missions; this became the Forestry Support Program (Steen 1991; USAID 2004a). In 1983 USAID issued policies on Environment and Natural Resources and on Forestry Policy and Programs, which, respectively, underscored its commitment to promote environmentally sound projects and sustainable use of forest resources (USAID 1983b, c). These established the policy framework for its environmental programs.

Reported funds to environment increased from US$13 million in FY 1978 to around US$220 million by FY 1986, and by the 1980s USAID's environmental program was expanding rapidly, fueled by the rise of the sustainable development discourse. USAID's FY 1982 congressional budget request argued that the agency's environmental objective was the "promotion of policies for sustainable development," and, illustrating the increasing influence of global environmental institutions, it cited as rationales the World Conservation Strategy, the Global 2000 Report, and the Interagency Report on Tropical Forests (USAID 1981). As sustainable development became the hegemonic discourse in environmental politics, being able to convey the economic importance of environmental issues was a critical means of integrating environment issues into USAID's development agenda. As a former senior USAID administrator recalled, "Articulating the rationale for environment as an economic issue was an important part of advancing the environmental agenda. We had to devise a rationale that was consistent with the agency's mission, and [US]AID's primary mission was economic development."[12] Likewise, in an early article about the emerging USAID environmental portfolio, the former USAID environmental advisor Molly Kux highlighted seven "investment rationales" for environmental conservation: maintaining ecosystem services; addressing rural populations' economic aspirations; increasing nature-based tourism; protecting endangered species; investigating natural economic products; building on indigenous conservation; and promoting sustained yield harvesting (Kux 1991). These early strategic framings aimed to appeal simultaneously to core advocates of economic development and species conservation. The resulting emphasis on the economic aspects of species conservation foreshadowed the rise of market-based conservation in the 1990s.

Concurrent with the ascendance of sustainable development, neoliberalism, with its emphasis on privatization and deregulation, was rising to prominence in mainstream economic policy in the 1980s. The Reagan administration specifically moved USAID away from a basic human needs orientation and state-led development and instead promoted greater American private sector involvement in foreign aid and private enterprise development in countries that received aid. Security related aid also assumed primary consideration (Auer 1998; Berríos 2000; Essex 2013). Nonetheless, the push by the Reagan administration to contract government services out to private companies triggered a reactionary move by the Democratic Congress to promote NGOs and to

protect environmental programs. At the same time, public donations to the membership-based environmental NGOs soared in the
1980s in response to Reagan's environmental rollbacks, and environmental NGOs developed networks like the Green Group, in
which the leaders of the largest U.S.-based environmental NGOs
coordinated their campaigns (Bramble and Porter 1992).

Organizing initially around the idea of sustainable development
in 1981, a group of NGOs formed the Global Tomorrow Coalition
(GTC) in an effort to stem a feared rollback of environmental programs by the Reagan administration and to build public and congressional support for global environmental issues (GTC 1981).[13]
Led by Thomas Stoel of NRDC and Russell Peterson, the president of the National Audubon Society and a prominent Republican,
the coalition included environmental NGOs as well as many of the
major development NGOs, such as Cooperative for Assistance and
Relief Everywhere (CARE) and Bread for the World.

Many of the active environmental NGOs based in Washington
at the time, such as the National Wildlife Federation, WWF-U.S.,
and the National Audubon Society, drew their mandates from the
general public's interest in saving animals. In reaction to the Reagan
administration, these species-focused groups helped to push through
FAA amendments that emphasized the need to protect endangered
species, forests, and biological diversity (Shaffer et al. 1987; U.S.
Congress 1983, 1986a; USAID 1988c).[14] A former congressional aide
recalled the activist bent of the group: "The NGOs were working
closely together. Not competing the way that they are today. . . . It
was an exciting time. We were going to change the world. We could
save the planet."[15] This close working relationship between congressional staff and NGOs established the precedent for future NGO–
congressional collaboration in shaping USAID environmental policy.

Growing data on forest loss and species extinction, above all in
the tropics, incited public concern about biological diversity in the
early to mid-1980s (Wilson 1988). The 1980 Global 2000 Report
by the U.S. Council on Environmental Quality and the U.S.
Department of State prompted government-wide awareness about
species loss (Shaffer et al. 1987). In 1981 Tom Lovejoy, then the vice
president at WWF-U.S., testified before a House Foreign Affairs
Authorization Committee on Energy, Environment, and Population

programs about the importance of protecting biological resources and linking conservation and development (U.S. Congress 1981b). That same year USAID and other government institutions sponsored a strategy conference on biological diversity, the recommendations from which were codified in FAA amendments in 1983. These amendments required USAID to take into account biological diversity in its activities and assigned the agency to organize an interagency task force to develop a strategy on biological diversity (USAID 1985). Members of Congress and their staffs were initially skeptical about the concept,[16] but, as a congressional aide recollected, the concept quickly took hold: "The first time I heard [the word] biodiversity I thought 'what the hell does that mean? What is biodiversity?' Within a year everybody was saying it."[17]

In 1985 the Subcommittee on Human Rights and International Organizations of the House Foreign Affairs Committee held a pivotal hearing to review whether USAID had followed the 1981 conference's biodiversity conservation strategy. Key NGO witnesses testified that USAID had tried to integrate conservation into existing programs rather than create new biodiversity-focused projects. Despite a plea made during the hearing by USAID Deputy Assistant Administrator for Science and Technology John Eriksson that biodiversity conservation should be integrated across all programs rather than isolated through an earmark (U.S. House of Representatives 1985), the FAA amendments of 1986 specifically set aside US$2.5 million for forests and biological diversity, underscored the need for "consultation with local people," and recommended that programs be implemented through NGOs (U.S. Congress 1986a; see also Shaffer 1987; USAID 1988). This authorization language established the foundation for funding the agency's biodiversity program with protected funds and through NGOs.

A Transforming Environmental Lobby

The 1986 amendments stated, "Whenever feasible, the president shall accomplish the objectives of this section through projects managed by private and voluntary organizations (PVOs)[18] or international, regional, or national nongovernmental organizations which are active in the region or country where the project is located" (U.S. Congress 1986a; USAID 1988c).[19] The preference for NGOs reflected an idealized vision of civil society as a counter to the private sector and cemented a close working relationship among NGOs, USAID, and congressional staff. A former congressional appropriations aide explained that the rationale for this language had been to address NGOs' complaints about contractors' failed attempts to implement environmental programs.[20] In interviews, a number of people complained about the idealized congressional perception of NGOs behind this differential access and resulting programmatic imbalance. As a former senior USAID official said, "If you're not for profit you have the moral high ground."[21] Likewise, a contractor retorted, "If AID finds out that one of their consulting firms is lobbying for an earmark you might as well kiss your next contract goodbye."[22]

The 1986 amendments inspired a USAID–NGO partnership—a former USAID environmental officer called it "a marriage of convenience for both USAID and the NGOs."[23] It prompted USAID not only to fund NGOs but also to train them to run USAID projects. USAID set up an office specifically to instruct NGOs on developing proposals for USAID grants.[24] It also prompted the formation, in 1987, of the Consultative Group on Biological Diversity, a consortium of private foundations interested in backing biodiversity conservation (Kux 1991; Shaffer et al. 1987; Vincent 1991). These mechanisms institutionalized coordination among the foundations, NGOs, and USAID. In 1985 the agency began awarding grants of around US$300,000 to the conservation organizations, primarily TNC and WWF-U.S. (Shaffer et al. 1987), including for the "Wildlands and Human Needs" program of the WWF-U.S. Ultimately, about half of the funds authorized by Congress for biodiversity went to NGOs and PVOs, and US$500,000 went to WWF-U.S. alone (Vincent 1991). A substantial portion funded the Biodiversity Support Program (BSP), which was designed to build USAID capacity in biodiversity and which ran from 1989 until 2001 through a noncompetitive grant totaling US$85 million to WWF-U.S., TNC, and WRI (BSP 2001). An intellectual force as well as a project manager, the BSP played a major role in shaping the future USAID biodiversity program and institutionalized close informal collaboration among the staff of NGOs, contractors, and USAID.

As the biodiversity program was developing, many of the Washington-based environmental advocacy groups like NRDC, which had previously advocated for USAID's environmental programs, were being drawn into fighting not only the rollback of domestic environmental programs by the Reagan administration (Bramble and Porter 1992; GTC 1981) but also the campaigns against the World Bank–sponsored Indian Narmada Dam and Brazilian Polonoroeste highway projects (Goldman 2005; Rich 1994; Wade 1997). An environmental advocate who had helped to create USAID's environmental program recalled, "We got to a point by the early eighties [in which] our ability to impact [USAID] from the outside had really diminished. When the Reagan administration basically came in and declared war on the environment . . . to the extent that there had been a focus in the environmental movement internationally all of the sudden it refocused domestically, to try and save the environment at home."[25] A former congressional aide similarly remembered that "a lot of the international people [from environmental NGOs] splintered off and formed the Bank Information Center . . . because they wanted to focus really heavily on the banks. USAID was secondary to them."[26] This latter campaign led to amendments that required the World Bank to conduct environmental impact studies and fund environmental programs, mandated that USAID establish an "early warning system" for IFI projects with potential environmental consequences, and obligated the American executive directors at the World Bank and the regional development banks to abstain or vote against projects that had not undergone environmental assessments (Aufderheide and Rich 1988; Bowles and Kormos 1995; Bramble and Porter 1992; Kormos, Grosko, and Mittermeier 2001; USAID 1988c).

At this critical turning point, the advocacy alliance behind USAID environmental programs transformed from an initial group of activists pushing a broad-based environmental agenda into an alliance of USAID grantees with a specific programmatic interest. As previous advocates for USAID's environmental programs turned their attention to the environmental consequences of the World Bank's programs and to the Reagan reforms, they left the wildlife conservation NGOs to attend to USAID. As grantees of the USAID biodiversity program, these organizations had a vested interest in protecting it.

This transition coincided with a critical shift in congressional practices. In the mid-1980s the authorizations committee began failing to pass the annual foreign aid *authorization* bill (Fleck and Kilby 2001; Heniff Jr. 2012; Lancaster 1999). As a result, the *appropriations* committee—which must pass annual legislation in order to keep the government working (or else pass continuing resolutions)—took over congressional foreign aid policy making, and the NGOs turned to the appropriators to leverage action.[27] This meant they were pushing not just for policy changes but also for financial commitments. Where previous directives by the authorizing committees had *authorized* agency biodiversity programs, the Appropriations Committee directed USAID *spending* on biodiversity conservation. In FY 1987 the Appropriations Committee directed USAID to spend US$4 million annually on biodiversity programs (U.S. Congress 1986b), and, by FY 1995, the committee was mandating that USAID spend US$25 million on biodiversity conservation. USAID expenditures on biodiversity increased five-fold between FY 1988 and FY 1991 (USAID 1991). As a USAID official recollected, "We couldn't scale up fast enough. In 1987 it was US$2.5 million, then US$5, then US$10 to US$15 million. Then we began exceeding the earmarks in the 1990s."[28]

Importantly, however, the agency's overall environmental program was also expanding. By 1991 the agency reported spending US$408 million on environmental concerns (USAID 1991). The administration of George H. W. Bush established a broad environmental agenda, one which focused on addressing five core areas: the "loss of tropical forests and other habitats critical for biological diversity; urban and industrial pollution; degradation and depletion of water and coastal resources; environmentally unsound energy production and use; and unsustainable agricultural practices" (USAID 1992b).

Reducing the State

Many of the NGOs that had organized in opposition to Reagan's antienvironmental policies came together again around the preparations for the Earth Summit in 1992. Environmental activists had high hopes that the 1990s would be the environmental decade

(Bramble and Porter 1992). One of the original environmental proponents who had worked on USAID's initial environmental programs recalled, "I remember going to the Earth Summit in '92 and being very proud of the seeds I had planted. USAID put together this glossy brochure of all these things it was doing on the environment, and I met many environmentalists from other countries who had gotten their start with USAID funding."[29] After hundreds of civil society organizations registered for the Earth Summit, the UN established a formal "Major Groups" structure to facilitate participation in the postsummit CSD process (Chatterjee and Finger 1994; Corson et al. 2015). These mechanisms institutionalized state–NGO collaborations similar to those coalescing in Washington.

In the wake of the Earth Summit of 1992 and at the end of the Cold War, the Clinton administration needed to find a new role for foreign aid. It sanctioned the use of foreign aid to address the environmental problems highlighted by the Earth Summit, including climate change and biodiversity conservation (Lancaster 2007). In budget justifications to Congress, the agency began advocating the importance of protecting the global commons and highlighting their strategic importance to Americans (USAID 1996, 1997a). Clinton appointees at USAID created a "global bureau," with four assistant administrators of sustainable development, one of which was focused on environment (Lancaster 1999). Within its Environment Center were three offices: environment and natural resources; environment and urban programs; and energy, environment, and technology (USAID 1998a). Many USAID environmental officers looked back with fondness at the Clinton years: as a former Clinton appointee recounted, "We basically had a White House–mandated pot of unearmarked, unprogrammed money that USAID got to spend on global environmental purposes." Within this broader environmental portfolio, biodiversity conservation and climate change held special positions. The appointee continued, "We believed they were the two most important global threats to the global environment—in other words, not threats to a particular country, but threats to the world as a whole."[30]

Nonetheless, the environmental enthusiasm brought on by the Clinton administration met with dramatic reductions in financial and personnel resources under both Clinton administration and Republican congressional neoliberal reforms. As part of the Republican congressional revolution in 1994 and the subsequent move to shrink the federal deficit, Congress dramatically reduced discretionary programs, including USAID's. The wave of conservative Republicans that swept into Congress that year put critical opponents of foreign aid into key committee positions. Sonny Callahan (R-AL), who campaigned on promises to limit foreign aid, headed the House Appropriations Foreign Operations Subcommittee, and Jesse Helms (R-NC) chaired the Senate Foreign Relations Committee. In 1995 Senator Helms launched a crusade against foreign aid, specifically accusing USAID of mismanagement, waste, and corruption and advocating the agency's reabsorption into the State Department. As a result of a deal between the White House and Helms, the USAID administrator began reporting directly to the State Department, laying the foundation for future reforms that more closely connected foreign policy, national security, and foreign aid. Second, as part of the Republican revolution and the move to reduce the deficit, discretionary programs, including USAID's, were slashed. U.S. foreign aid fell from over US$12 billion in 1993 to US$9 billion in 1996 (Auer 1998; Lancaster 2007).

Ironically, as Congress reduced foreign aid spending, it simultaneously increased its micromanagement of USAID's activities through legislative directives (Smillie et al. 1999; Zeller 2004). The former USAID deputy administrator Carol Lancaster (1999) argues that "by the 1990s, the United States no longer had a 'development paradigm' around which to shape its aid policies. Rather it had a set of activities it was prepared to fund, and those activities often arose from political pressures from NGOs and restrictions from Congress" (100). In the environment sector there were, by the mid-1990s, congressional directives for saving African elephants, global warming, renewable energy, and biodiversity, a situation which limited the agency's ability to choose its own priorities (Lancaster 1999).

Concurrently, the Clinton administration continued previous Republican administrations' neoliberal efforts to reduce government and to contract out public programs to private companies and NGOs. Further, it transitioned USAID from "an agency of

U.S. direct-hires that largely provided direct, hands-on implementation of development projects to one that manages and oversees the activities of contractors and grantees" (GAO 2003, 6). By 1997 USAID had closed more than two dozen overseas missions and through attrition and reduction in forces had reduced its staff by one-third (Berríos 2000; GAO 2003; Lancaster 2007; Smillie et al. 1999). As the agency eliminated permanent slots, it began hiring temporary contractors overseas and in Washington to do the work permanent employees had previously done (Zeller 2004). One longtime USAID official stated, "When I joined [USAID], it had sixteen thousand employees.... Now it has sixteen hundred, maybe two thousand."[31] As the agency reduced the ratio of staff to funds, grant sizes increased, and staff became less involved with the technical aspects of projects, turning them over to contractors and grantees instead. "Instead of one manager to five people," a USAID official summarized, "it became one to ten, for example, [and staff] had to move out more money per person. A typical AID program was no longer [US]$2.5 million, but a [US]$25 million project."[32]

The privatization reforms converged with the Clinton administration's vision of environmental NGOs as key constituents (Lancaster 1999). In the words of a former USAID senior official, "Al Gore in particular believed, [and] certainly [former USAID administrator] Brian Atwood perceived, that the NGOs were one of the key constituencies for USAID. It was groups like WWF-U.S., CI, and TNC whose priority was biodiversity, and it was groups like NRDC and others who were focused on climate change."[33] USAID's National Performance Review report—which responded to Vice President Gore's effort to make government more efficient, effective, and businesslike (Gore 1993)—specifically encouraged funding U.S. and foreign NGOs (Atwood 1993; Corneille and Shiffman 2004; Gore 1993). USAID's New Partnership Initiative in 1995 underscored engaging civil society and the private sector in USAID projects, and it required the agency to channel 40 percent of aid funds through NGOs (Esman 2003; USAID 1995a). By FY 2000 USAID was directing about US$4 billion of its US$7.2 billion program to NGOs (GAO 2002).[34] Senate appropriators continued to mandate that USAID fund NGOs to carry out the biodiversity program, even as the agency was downsizing. The Senate appropriations report for FY 1996 stated, "As USAID makes efforts to downsize, it should remain active in regions that are significant for global biodiversity . . . [and] NGOs are often the most cost-effective channels for delivering development assistance" (U.S. Congress 1995).

The combination of the downsizing and congressional backing of biodiversity meant that USAID was reducing its staff just as the Congress was requiring it to spend increasing amounts on biodiversity.[35] As a result, "[USAID] never really staffed up to meet the increasing environment money and extra funds that were being forced down our throats."[36] Instead, the agency directed increased funding to the emergent conservation organizations.

In the 1990s the conservation NGOs experienced tremendous growth. Investments by WWF-U.S., TNC, and CI in conservation in the developing world grew from roughly US$240 million in 1998 to close to US$490 million in 2002 (Chapin 2004, 22). CI's "Hot Spots" strategy accompanied an increase in overall annual expenditures from US$27.8 million in 1998 to US$89.3 million by 2004, and WWF-U.S.'s "Ecoregions" program accompanied a rise in expenditures from US$80 million to US$121.7 million between 1997 and 2005 (Rodríguez et al. 2007). USAID was a contributor to this growth: "USAID provided a total of roughly US$270 million to NGOs, universities, and private institutions for conservation activities. The lion's share of this amount destined for NGOs was harvested by WWF[-U.S.], which received approximately 45 percent of the available money" (Chapin 2004, 24). "The five largest conservation organizations, CI, TNC, and WWF[-U.S.] among them, absorbed over 70 percent of that expenditure" (Dowie 2005, 4). Conservation NGOs also started pursuing corporate partnerships in the 1990s, and as they grew they became more influential, not just with respect to USAID but also more broadly in Washington politics.

Thus, by the turn of the century USAID had shrunk, turning over much of its project management activities to contractors and grantees, while conservation organizations had grown, fueled in part by Clinton administration and congressional policies that encouraged the funding of NGOs to carry out the growing biodiversity program. In the second half of the 1990s, as part of a Congress-wide effort to reduce earmarks, the Appropriations Committee stopped mandating specific amounts for biodiversity, instead simply reiterating each year the need to fund conservation. With a vested interest in USAID's biodiversity program and concerned that the agency was using biodiversity funding for other aid programs, the conservation NGOs, with the backing of selected USAID staff, convinced Congress to reinstate mandates for biodiversity expenditures.[37] These directed "AID to restore overall biodiversity funding as well as funding to the Office of Environment and Natural Resources to levels that reflect the proportion of funding of development assistance provided in fiscal year 1995" (U.S. Congress 1999).

Narrowing the Environmental Portfolio

The George W. Bush administration brought various changes that reversed many of the environmental gains accomplished under Clinton. Bush's election in 2000 marked the ascension to power of a neoconservative administration that, in the wake of the destruction of the World Trade Center in September 2001, explicitly blurred the lines between foreign aid and national security policy (Essex 2013; Hills 2006). While funds for foreign aid rose "at one of the fastest rates in the history of U.S. aid-giving, expanding by roughly 40 percent between 2001 and 2005" (Lancaster 2007, 91), the environment was not a Bush priority. In a major overhaul of foreign aid, the administration created a new foreign aid agency, the Millennium Challenge Account (MCA), designed to assist countries that met U.S.-determined standards for governance and neoliberal economic reform (Chhotray and Hulme 2009; Mawdsley 2007; MCC 2007). Madagascar was the first country to meet the standards, submit a proposal, and receive funds from its funding arm, the Millennium Challenge Corporation (MCC), to the tune of US$110 million over four years, roughly equivalent to the USAID Madagascar budget over the same time frame.

Emphasizing the economic value of conservation became a survival strategy. The Bush administration instituted a foreign aid reform that aimed "to reduce waste and demonstrate responsible use of taxpayer" dollars (U.S. Department of State 2006) and, ironically, to convince Congress to reduce its earmarks in the foreign aid appropriations bill (USAID 2004b). The reform moved USAID's policy office to the State Department and classified all USAID programs into one of five categories: (1) peace and security; (2) governing justly and democratically; (3) investing in people; (4) economic growth; and (5) humanitarian assistance. The USAID administrator Andrew Natsios abolished Clinton's Global Bureau and subsumed environmental programs under a new Bureau for Economic Growth, Agriculture, and Trade, which prioritized agriculture, health, democracy, trade, and humanitarian assistance. Environmental programs were reorganized as a subcategory under economic growth, with specific emphases on biodiversity conservation, natural resources, and reduction of pollution (U.S. Department of State 2007b). Missions were required to decrease the number of Strategic Objectives (sectoral programs) they funded,[38] and environmental programs often became subcomponents of other sectoral programs or were eliminated altogether.[39] Finally, through the Global Development Alliance, which aimed to leverage private sector funding for development, USAID became a linchpin in the expanding NGO–corporate conservation partnerships. Among these were, for example, the Sustainable Forest Products Global Alliance, a US$23 million initiative among Home Depot, Metafore, the U.S. Forest Service, and WWF-U.S. This relegation of environment to the economic growth portfolio and emphasis on public–private partnerships reinforced the importance of articulating environmental programs in terms of their contribution to economic growth and outreach to the private sector.

In contrast to the Clinton years, during the Bush era overarching foreign aid policy virtually ignored environmental concerns (USAID 2004b), and the biodiversity earmark became one of the few means to fund environmental programs. A former USAID environmental officer said, "We tried like hell [to integrate environment into the white paper]. . . . We presented a set of actions adapted to the different categories of the white paper showing that natural resources management [and] environmental issues were fundamental to achieving what USAID wanted to achieve in each of the categories of countries . . . but Natsios made it clear it wasn't of any importance to him."[40] Likewise, the MCA ignored environmental issues until conservation NGOs pushed Congress to require it to add an environmental indicator that included as one of four subindicators the protection of at least 10 percent of its biomes in national parks.[41] NGO lobbying on Capitol Hill became a critical means of protecting USAID's environmental programs. Asked to comment on the aid reforms, a former USAID environment officer lamented, "Environment is essentially a footnote. . . . And the funding is only there because of the biodiversity earmark."[42]

As happened in the Reagan years, the demotion of environmental issues revitalized NGOs. But by this point the primary advocates on the Hill for USAID's environmental programs were the conservation NGOs,[43] and they focused on biodiversity rather than on environmental issues broadly or on integrated conservation and development. A former USAID official recalled the challenge of building political support for addressing agricultural drivers of biodiversity loss: "[During the Clinton years] there was no champion on the Hill for increasing overall agricultural funding . . . [but] there was a very effective champion for biodiversity conservation."[44] Another USAID official complained about the same issue in the second Bush administration: "There's a huge constituency for pandas or gorillas, but there's no constituency for rural development. . . . There is no lobbying for sustainable use."[45] And a former USAID official confirmed, "WWF-[U.S.] and TNC are still there pushing very hard. But in terms of energy, climate change, pollution issues, that lobby has gone away."[46] The result of this transformation in the entities that lobbied for USAID's environmental programs was a narrowing of the agency's environmental priorities around the special interests of its political supporters.

USAID staff pursued creative ways to continue funding a broad array of environmental programs. In 2002 staff members in the Africa Bureau in USAID's headquarters, together with individuals from WRI, Winrock International, and International Resources Group (IRG), published an innovative paper entitled "Nature, Wealth, and Power" (NWP), which argued that environmental programs that recognized governance, economic, and social aspects of environmental issues were more successful than those that did not (Anderson et al. 2002). One of the authors explained the paper's purpose: "The problem with environment now is that it has been labeled and narrowly defined. It is pejorative within the White House and the agency. People think that environment has nothing to do with conflict, governance, or economic growth. 'Nature, Wealth and Power' was an attempt to show these linkages and explain how rights to resources underlie peoples' livelihoods."[47]

While NWP sought to counter the isolation of the environment program and to promote cross-sectoral linkages, legislative and bureaucratic impediments prevented cross-sectoral coordination, and many interviewees focused on the tension between agriculture and environment programs. Many of the environmental officers had come in with the Clinton–Gore administration and had previous NGO backgrounds or ties with Capitol Hill or both, but agriculture officers were often former Peace Corps volunteers and field oriented: "You had this whole crew of 'aggies' who came out of Texas and the Midwest, and they all ended up working in international development."[48] Interviewees cited a "cultural difference" between the USAID agriculture and environment officers: "I think that there's a different mindset between people who think that biodiversity is the most important thing and those who think rural poverty alleviation is the important thing. The environmental constituency is much more zealous. It doesn't seem to be as introspective."[49]

These cultural differences were reflected in bureaucratic processes and funding competition. Environment programs were typically managed through cooperative agreements, in which USAID would buy into an existing NGO program or through a grant, agree on the objectives, and then give the NGOs leeway to implement the program. In contrast, agriculture programs were often managed by contractors, in which USAID procures goods and services—often from a for-profit organization—and then monitors the achievement of particular deliverables.[50] In the 1990s, as environmental funding skyrocketed, funding for agricultural projects declined (Hoben 1997). A USAID official recounted that during the buildup of the environment program "there was lots of resistance to integrating agriculture and environment because of competition for [the] same funding."[51] A former senior USAID official confirmed that "in the early nineties there was a direct correlation between environment funds going up and agriculture funds going down."[52] Thus, when Natsios came in with a strong commitment to agriculture, congressional members reacted by introducing legislation that protected forests, biodiversity, and wildlife from any associated cuts.[53]

As USAID staff tried to create ways to use biodiversity funds for a range of environmental programs, the conservation advocates pushed back, trying to protect USAID's biodiversity funds from being used by other USAID environmental programs. They lobbied Congress to ensure that biodiversity funds were not used for non-conservation activities. Beginning in FY 2002 the Appropriations Committee began including legally binding biodiversity conservation earmarks in the appropriations bill itself rather than just in the accompanying report language, where directives are technically not binding (although in practice they are treated the same). NGO, USAID, and congressional staff drew up strict guidelines that overseas missions had to follow in order to use these biodiversity-earmarked funds (see chapter 6). The combination of pressure to fund biodiversity and stagnant overall environment funding forced program managers to try to conform their environmental projects to the biodiversity funding criteria in order to qualify for biodiversity funds (Hecht, Gibson, and App 2008).[54] As one agency official explained, "Programs that really should be doing natural resource management have had to shift where they work to be closer to an area of biological significance, or they have had to add a new dimension that looks specifically at biodiversity."[55]

By the turn of the century, biodiversity, as one of the few protected environment programs, began to absorb the majority of USAID's environmental funds, redefining what constituted the environment within USAID's portfolio. One USAID official commented, "With Bush and Cheney, environmental funding is down, but Congress is still earmarking, so 90 percent of our activities are biodiversity conservation [and] they have to meet those [biodiversity] criteria."[56] In FY 2006 the US$165.5 million set-aside absorbed the majority of environment funds, which was just shy of US$200 million that year. As another official said, "[As] the biodiversity conservation earmark rose, environmental funding was frozen and then it started going down. It literally forced the end of nonbiodiversity environmental work."[57]

In contrast to their predecessors twenty years earlier, who had campaigned for USAID funds for managing the environment so as to ensure rural livelihoods and sustainable economic growth, conservation NGOs pushed to exclude such programs. In congressional testimony they expressed their concerns that congressionally earmarked funds had been used to fund "natural resource management programs," which "[include] a multitude of activities, many of them extractive in nature, that are not necessarily consistent with species conservation or the protection of ecologically important habitat" (Patlis 2007). Similarly, Washington-based USAID staff—called by opponents of the process the Washington biodiversity police—began to visit missions and review mission requests to approve or disapprove of programs for biodiversity funding against these criteria. One NGO lobbyist explained the need for the guidelines: "In the field, there's no agreed-upon definition of what is a biodiversity project. So USAID is funding, for example, forestry-related projects that are really more about sustainable forestry, and they're calling them biodiversity, and they're claiming credit for this against the [earmark]. So the real conservation projects are under pressure."[58]

Faced with this political pressure, agency officials found it harder to finance programs that addressed broader cultural, economic, social, and political drivers of environmental change, including biodiversity loss.[59] Congressional aides argued that because "biodiversity gets at all environmental problems,"[60] the earmark could fund policy and drivers of biodiversity loss. Yet the guidelines clearly prevented the funding of problems like urban pollution and waste management, and, as one of the D.C.-based managers of USAID's biodiversity portfolio admitted, even conservation policy was hard to fund under the earmark guidelines: "Because policy is more broad-reaching than any area of biological significance, it's harder to figure out how to fund that."[61] This narrowing agenda undermined conservation, as programs that sought to address it more holistically or to combat the broader driving factors of biodiversity loss often did not qualify for funding.[62]

Creating the New Conservation Enterprise

In 2003 WWF-U.S., TNC, CI, and WCS created the ICP, which aimed to build congressional support specifically for international biodiversity conservation by organizing activities such as congressional briefings and lunches, jointly endorsed letters, and overseas congressional trips to priority biodiversity sites.[63] One of the ICP's primary activities was the annual publication of an International Conservation Budget (ICB), which recommended appropriations levels for the major U.S. government–funded international biodiversity conservation programs, including USAID's. Its success is revealed by the fact that the amounts legislated in the appropriations bills each year often mirrored those promoted in the ICB. For example, for FY 2008 the ICB recommended US$195 million for USAID, which was the amount that the Senate Appropriations Committee included in report language later that year (U.S. Congress 2007).[64] In 2003 the ICP inspired the creation of a bipartisan House International Conservation Caucus (ICC), followed in 2005 by a parallel caucus in the Senate.[65] Composed of an eclectic bipartisan membership ranging from the far left to the far right, the ICC had 150 members by the end of 2007 and had become one of the largest bipartisan caucuses in the House. The strict focus on *foreign* environmental issues allowed its leaders to craft a bipartisan coalition that included a broad spectrum of political persuasions[66] and to draw on reliable public concern in the United States about—and therefore congressional interest in— saving charismatic megafauna in other countries. It also permitted members of Congress to embrace environmentalism without confronting their domestic constituents. The former USAID official quoted in the epigraph above went on to say, "When there are problems with local communities [overseas], they don't call up their congressman."[67] Accordingly, the caucuses attracted individuals who considered themselves antienvironmentalist on domestic issues by offering a way, according to a congressional aide, for them "to be proactive when it comes to the environment without being labeled a traditional environmentalist."[68] These diverse individuals came together, as an NGO congressional liaison summarized, because "they [the members] all like wildlife, and they have all at one time or another visited international park sites abroad."[69] In this respect, NGO-organized trips for congressional members and staff to biodiversity sites overseas were important mechanisms for building congressional interest in funding international biodiversity conservation;[70] as another aide observed, "In his/her travels [name of Congressman/woman] sees so many different examples of people not taking care of natural resources effectively."[71] Most congressional staff I interviewed had been on overseas jaunts with one or more of the four ICP partners. However, these trips also reinforced the idea that conservation was a foreign problem, thereby eliding any American government role in promoting drivers of domestic or foreign environmental issues.

In July 2006 the ICP formed the ICCF, a separate 501c(3) organization whose mission was to assist the ICC and specifically to create "an educational forum on Capitol Hill, where we keep Members of Congress and their staff constantly updated with information we synthesize from our base of NGO supporters on the most pressing and timely issues in international conservation" (ICCF 2007b).[72] For example, the ICCF presented congressional briefings on such topics as Walmart's commitment to sustainability, the USAID-funded program Living in a Finite Environment in Namibia, and the ecosystem payments program in Costa Rica.

Drawing on the corporate linkages of some of its founding NGOs, the members of the ICCF's advisory "conservation council" included corporate giants like ExxonMobil, International Paper, and Unilever. TNC's corporate associates and major contributors at various times comprised 3M, Shell Oil, General Motors, Ford Motor Company, BP Exploration, MCI Telecommunications Company, MBNA America Bank, Enron Corporation, Georgia-Pacific, Johnson and Johnson, Weyerhaeuser Company, Waste Management Inc., Monsanto Company, and Dow Chemical (Bailey 2006); "some 1,900 corporate sponsors" donated a total of US$225 million to TNC in 2002 (Chapin 2004, 24); and "CI's website lists over 250 corporations, which donated approximately US$9 million to its operations in 2003" (Chapin 2004, 24). In 2008 these corporations included, among others, Anglo-American, Chevron, and Rio Tinto (CI 2008). An interviewee from the ICCF justified these partnerships: "Environmentalists out of principle tend to shun people like ExxonMobil and Walmarts of the world. But we need their support. They're an economic gorilla, and they're going to be here, so you can either engage them or you can alienate them."[73]

ICCF's "Partners in Conservation" brochure showcased a number of public–private partnerships undertaken by the organization's sponsors (ICCF 2007a). For example, it cited the Walmart and the National Fish and Wildlife Foundation Acres for America program, which conserved one acre of critical wildlife habitat for every acre of land developed for an existing Walmart facility or new one created in the United States, and ExxonMobil's investment in the Save the Tiger Fund, which, it boasted, "represents the largest single corporate commitment to saving a species" (ICCF 2007a).

As colorful collages of corporate and NGO logos, the widely circulated invitations to the ICCF's annual galas were striking symbols of the merging of conservation and capitalism. In 2006, for example, attendance at such fund-raising events cost between US$1,000 and US$50,000. They honored various celebrities, including the former United Kingdom prime minister Tony Blair, the actor Harrison Ford, and Chad Holliday, the chairman and CEO of DuPont for their contributions to international conservation. Concurrently, the ICCF drew on the American conservation roots in big game hunting: at its second gala in September 2007 the ICCF honored Blair with the 2007 ICCF Teddy Roosevelt International Conservation Award. The 2006 and 2007 invitations also boasted meals prepared by the "Texas Cowboy Chef Tom Perini," who was "the Caterer to the President of the United States." The galas represented vehicles for colossal shifts of funds among U.S.-based state, private, and nonprofit sectors in the name of foreign conservation and, as such, contributed to what I call a conservation enterprise, in which funds circulate among the world's elite without ever being used on the ground for conservation (Corson 2010).

In spite of the organization's efforts to increase congressional appropriations for biodiversity conservation, the ICCF invoked antigovernmental rhetoric to attract conservative and corporate members. The president of ICCF, David Barron, underscored the bipartisan nature of the foundation and its neoliberal tenets at the organization's inaugural gala in September 2006. In a published letter to the gala attendees, he stated, "We are *not* advocating more government. Quite the contrary, we are advocating private sector solutions. . . . *We are pro-development and pro-business. We are pro-people, pro-wildlife and pro-wilderness.*"

The ICCF went on to bolster similar work in congresses and parliaments around the world. Its outreach to conservative and

corporate leaders was a successful strategy to raise funds and awareness for environmental conservation. Similar to sustainable development, international biodiversity conservation became a nucleus
around which public and private organizations could find common
cause. By defining the environment as a foreign concern centered on
wildlife conservation, the ICCF's high-profile effort permitted organizations and individuals who were opposed to environmental issues
in general to become green. In doing so, the ICCF attracted a broad
spectrum of congressional and corporate support. However, in the
process it offered a green stamp of approval for corporate partners
to continue their exploitative practices, and it created avenues for ex-
orbitant expenditures of wealth in the name of conservation, without
translating that mobilization into conservation on-the-ground.

Reconfiguring Environmental Governance

While in-depth analysis of dynamics in the Obama administration
is beyond the scope of this book, I will highlight two events. First,
in 2009 the Senate appropriators began directing USAID, with the
Departments of State and Treasury, to spend US$1.2 billion on en-
vironment and climate change. This directive was an effort to
bring climate change and biodiversity directives under a broader,
protected rubric of environmental programs more generally.
Second, in 2014 the agency released what it termed its First
Biodiversity Policy, which aimed to integrate biodiversity more co-
herently into the agency's overall development mission and to "re-
commit" the agency "to conserve biodiversity through strategic
actions to reduce threats and drivers, as well as a new focus on integrating biodiversity conservation with other development sectors." One of its stated goals was to "allow management units to
better justify working on some of the key drivers of biodiversity
loss, in addition to the immediate threats" by requiring missions to
conduct analyses of drivers and threats with an eye toward development theories of change (USAID 2014a, 21). This may indicate
a shift away from the strict interpretations that have undermined
both the ability to address broader drivers of biodiversity loss and
other environmental issues. However, understanding the politics
behind and impact of these changes requires further study.

In this chapter I have argued that the reduction of the state under neoliberalism, which opened up room for private actors to influence state policy, enabled the formation of a dynamic alliance among members of Congress, USAID, an evolving group of environmental NGOs, and the corporate sector around biodiversity conservation funding. In the emergent forms of environmental governance, nonelected agents—from both the not-for-profit and the private sectors—increasingly shaped public policy, and idealized visions of NGOs as civil society and a countering force to corporations underpinned their influence, despite their contemporary corporate partnerships. As the state turned to private organizations to implement its work, it became dependent on these entities not just to design and implement programs but also to mobilize political support to sustain them.

As the assemblage of NGOs pushing environmental foreign aid evolved, so too did the definitions of "environment" encompassed in congressional requirements for USAID's programs. USAID's environmental program began in the 1970s with a focus on emphasizing natural resource management for livelihoods and economic growth. In the mid-1980s, as a result of the advocacy efforts by conservation NGO representatives, members of Congress, and USAID staff, Congress began earmarking funds specifically for biodiversity. As overall environmental funds became more limited during the second Bush administration, these congressional requirements and others overwhelmed USAID's environmental programs, forcing a reduction in funding for other environmental issues. At the same time, strict guidelines on the use of biodiversity funds made it harder for USAID staff to fund cross-sectoral environmental projects as well as projects that addressed the broader political, economic, and social drivers of environmental degradation, including those of biodiversity loss. The result was an increasing prioritization within USAID's overall environmental program on the special interests of its political champions.

In direct contrast to 1970s environmentalism, which had challenged the idea of infinite economic growth expansion and promoted state-led policy, the ICCF built its arguments and legitimacy around a neoliberal conception of governance, arguing for the reduction of government and increasing engagement of the private sector, even as it endeavored to influence state funding. Furthermore, it focused strictly on foreign biodiversity conservation, in the process enabling supporters to claim environmental credentials without having to address difficult domestic issues or their own companies' exploitative practices. In this manner ICCF facilitated capital accumulation not only by offering corporations a stamp of approval for their environmental stewardship but also by fueling the growing capitalist enterprise forming around the concept of biodiversity conservation. While this politically successful mobilization of congressional interest in biodiversity conservation has brought new funds and political endorsements, it has also impeded the agency's ability to develop a broader environmental program and to effectively address the drivers of biodiversity loss—an irony I explore further in chapter 6.

CHAPTER FIVE

A Model for Greening Development

I come from that island which astronauts recognize, from the height of their space craft, as "a spreading bloodstain on the blue of the Indian Ocean." . . . Shortly after our independence, toward the end of the 60s, our leaders reacted [list of environment conferences and policies]. . . . Unfortunately, these policies, which are carried out with the agreement of traditional external aid donors, are mainly directed toward the priority of saving our heritage of biodiversity. The policies of conservation have indeed been useful in the face of the degeneration of our environment and the extreme vulnerability of our ecosystems. Nevertheless, they have not [addressed] the human aspect. . . . We can do nothing about this, for lack of means. Madagascar can no longer sustain the weight of its external debt, which holds back all economic and social development.

Further, among those debts are those due for saving of our
environment.

—PRIME MINISTER OF MADAGASCAR GUY WILLY RAZANAMASY,
United Nations Conference on Environment and
Development, Rio de Janeiro, 1992

THE UNITED NATIONS CONFERENCE on Environment and
Development held in Rio de Janeiro in 1992 was a pivotal moment in
global environmental politics for a number of reasons. It led to binding international environmental conventions; it consolidated state authority in multilateral agreements; and it institutionalized long-term
participatory mechanisms for including nonstate actors in global environmental politics. Nevertheless, many southern governments were
handicapped in their ability to pursue the idealistic initiatives that
emerged from it. The epigraph above illustrates the bind in which
the Madagascar government found itself. Facing a heavy debt burden,
the government signed its first structural adjustment agreement with
the IMF in 1980, and subsequent reforms mandated a reduction in
state programs and the pursuit of rapid economic growth through
privatization and liberalization. While the government's acquiescence
to structural adjustment prompted an influx of bilateral and multilateral environmental aid, the launch of Madagascar's NEAP at the particular historical moment of the late 1980s and early 1990s shaped
the realm of possibilities for its subsequent agenda. As donors and the
state were creating new environmental programs, the government
was liberalizing the economy, offering incentives for natural resource
exploitation, and reducing state management and enforcement capacity. Concomitantly, the World Bank was promoting the involvement
of civil society in public policy-making processes. Numerous nonstate
actors developed and implemented the NEAP, and its final priorities
reflected their interests and agendas.

In consonance with Mosse's (2005) proposition, the NEAP's
greatest success was its ability to mobilize and stabilize critical political relationships among donors and nonstate actors. It brought
together and aligned a diverse set of state and nonstate actors
behind a coherent environmental agenda. It offered an "example of

how the NEAP process can be used to create synergy between do-nors" (Greve, Lampietti, and Falloux 1995, 7). As a former USAID official summarized, "Madagascar is one of the only examples I can think of where you have a fifteen- to twenty-year goal with multiple donors all buying into that goal. It may not be perfect, but I can't think of any place else that has done that."[1] Many attributed its achievements—which they listed as the creation of new environ-mental agencies and laws, increased government leadership in envi-ronmental issues, the building of regional capacity in environmental management, and the expansion of the parks network—to the long-term commitment of a core group of scientists, donor officials, con-servation NGO leaders, and Madagascar government officials.[2]

Nonetheless, the need to sustain the relationships among these actors undermined the program's ability to pursue critical objec-tives, as over time the NEAP's priorities shifted to mirror those that they would all endorse. On paper the NEAP proposed a broad environmental program, and numerous individuals underscored the need to consider macroeconomic policy together with environ-mental planning, to integrate rural development and environment, and to ensure a broad-based environmental program that ad-dressed the ultimate drivers of land use change. However, three is-sues undermined the pursuit of these goals. First, disproportionate funding for biodiversity—especially from U.S.-based donors and NGOs—sidelined other priority issues, such as urban pollution and environmental education. Second, rather than tackling govern-ment capacity to regulate natural resource–based industries and ensuring the incorporation of environmental concerns into eco-nomic planning, many programs prioritized the redressing of rural peasant land use. A third issue was that as donors leveraged the creation of new, independent institutions in order to circumvent weak, corrupt government agencies, they reinforced the state's de-pendence on external actors for program design, management, en-forcement, and funds. In doing so, the environmental program evolved in ways that left the country vulnerable to expanding com-mercial resource extraction in the twenty-first century.

In this chapter I analyze how state and nonstate actors negoti-ated programmatic priorities during three phases of the NEAP. I concentrate on how and why biodiversity conservation, achieved through the expansion of protected areas, came to dominate much of Madagascar's donor-funded environmental program, marginalizing initiatives like sustainable agriculture and rural development, CBNRM, state environmental management capacity building, and green economic planning. I argue that biodiversity conservation became, with strong backing from the United States, an avenue through the tensions and contradictions of trying to create an environmental agenda in a neoliberal political-economic context. It comprised a political pathway that ensured the environmental agenda did not impede rapid economic growth.

In focusing my analysis on these issues, I do not question the assumptions made by the creators of the NEAP about the causes and extent of environmental degradation in Madagascar. A number of other scholars have already done this (e.g., Bertrand and Ratsimbarison 2004; Jarosz 1993; Klein 2004; Kull 2004; McConnell and Kull 2014). Similarly, I do not join the ample list of consultants and academics who have evaluated the on-the-ground effectiveness of NEAP programs (e.g., Brinkerhoff 1996; Gezon 1997; Hough 1994; Lindemann 2004; Marcus 2001; McCoy and Razafindrainibe 1997; Peters 1998; Swanson 1997). Rather, I concentrate on the political dynamics that shaped the NEAP, and, as in other chapters, my analysis is U.S.-biased. I begin by discussing the neoliberal context in which the NEAP came into being. Then, I review how programmatic priorities changed over each of its three phases. I show how national government policy sought to balance pressures to implement neoliberal economic policy and conservation pressure by embracing the expansion of protected areas as a pathway through their contradictions. Finally, I discuss the threats posed by the rapidly expanding mining and illegal forestry industries and the lack of government capacity to control them. In the next chapter I examine how the political dynamics between the U.S. government and U.S.-based conservation NGOs shaped USAID's contributions to the NEAP and the rise of biodiversity conservation as a priority within it.

Foreign Aid for Green Neoliberalism

By 1989 1 billion of Madagascar's US$2.6 billion debt was owed to the World Bank (Jolly 2015), and although the presence of environmental aid may have provided some leverage for the Madagascar government in its negotiations with the IFIs (Barrett 1994), the World Bank had substantial influence over the country's economic policy. Over 75 percent of the US$96 million that the World Bank gave between 1981 and 2005 was directed toward formulating a liberal regulatory framework (Sarrasin 2003, 2007a), specifically to "accelerate export-led growth by increasing private investment and productivity through reforms in the policy and business environment, upgrading of private firms' capabilities and global market knowledge and involvement, and attracting FDI" (World Bank 2001, n.p.). Thus it is not surprising that the Malagasy Environmental Charter of 1990 (Law 90–033), which codified the NEAP, argued that "structural adjustment in management of our economy should be linked to adjustment of our management of natural resources" (Democratic Republic of Madagascar 1990, 20).

The environmental charter specifically called for the reduction of the state, liberalization of the economy, and involvement of non-state actors in environmental policy making. It attributed environmental degradation to state interference, and it argued that "the failure to adequately decentralize power, combined with decreasing buying power among the populace, has led to the relentless onslaught of corrupt practices at all levels throughout the country. Moreover, excessive state control of the economic factors of production has led to a total loss of private individual initiative, the main force behind economic development that the country experienced previously" (Democratic Republic of Madagascar 1990, 14). In order to combat this "failure" and to stimulate private resource conservation, the charter emphasized the need for "less administration, but better administration" (Democratic Republic of Madagascar 1990, 20). It promoted the withdrawal of the state from productive activities and the decentralization of decision making, organization, and budgets: "Because the environment is everyone's business, the government should yield its place to private operators. The government role should be to set policy, develop needed incentives, and monitor and evaluate actions in the field. . . . This means stimulating associations of users, the NGOs, and private enterprises who are attempting to conserve and increase the value of the country's resources. This long-term process should get the maximum number of actors involved in conservation of the environment" (Democratic Republic of Madagascar 1990, 23).

The law passing the charter then laid out more clearly the participatory nature of governance that was envisioned: "Environment management is jointly carried out by the State, decentralized authorities, nongovernmental organizations, economic operators, as well as all the citizens" (République Démocratique de Madagascar 1990, 2, author's translation). It emphasized that the environment is "everyone's concern," and it presented the charter as a social contract—a reference for a *sensibilisation*[3] campaign at the end of which environmental concerns should comprise a part of the everyday actions of local administration and all citizens. Redressing environmental degradation, it contended, would entail creating incentives for citizens and businesses to conserve resources and making the poor aware of the negative consequences of their actions. Finally, the law emphasized the need to develop programs that would benefit local communities and to solicit community input in program development. It underscored that the third and final phase of the NEAP "should be a period of 'withering away' of environmental institutions, after which the environment and its concerns should make up a part of the everyday administration of local authorities and every citizen" (Democratic Republic of Madagascar 1990, 45; see also Falloux and Talbot 1993; Hufty and Muttenzer 2002; Sarrasin 2007a, b). In this manner the environmental charter placed the responsibility for change in the hands of the individual, reduced government capacity, and elided how to address the environmental impacts of liberalized trade.

The NEAP's implementation, like its creation, entailed extensive consultations with nonstate actors. Its three phases were originally scheduled as the following: environment program first phase (EP1), 1991–95; second phase (EP2), 1996–2000; and third phase (EP3), 2001–5, although the schedule slipped slightly to EP1 (1991–97), EP2 (1997–2003), and EP3 (2003–8) because of the political turmoil in Madagascar in 1991–92 and 2001–2.[4] A coordinating committee composed of all the major donors and executing agencies and ministries met annually, and its meetings were attended by representatives of donors, foreign and local NGOs, private sector organizations, academic institutions, government ministries, and parastatal agencies, including newly created executing agencies of the environment program. Many of the attendees were foreigners (Brinkerhoff and Yeager 1993; Froger and Andriamahefazafy 2003; Greve 1991b, 1992, 1993).

The state depended on foreign aid to implement the environment program, and as a result donors had considerable influence over its priorities. Between the early 1990s and 2009, foreign donors and the Madagascar government together provided the equivalent of almost US$500 million for the environment program, including an estimated US$150 million for EP1, US$150 million for EP2, and US$170 million for EP3 (Razafindralambo 2005; World Bank 1997, 2003c). The government's contribution ranged from 9 to 20 percent of costs, and it peaked in EP2, almost doubling in comparison to EP1. However, because donors' contributions increased as well, the government contribution dropped to less than 9 percent of the NEAP costs in EP3 (Horning 2008b). Of the original estimated costs for the NEAP, 33 percent was to go to biodiversity protection, 20 percent to mapping and geographic information systems, 15 percent to soil conservation and watershed miniprojects, 11 percent to land titling, 11 percent to institutional policy support, 6 percent to education and training, and 4 percent to environmental research (Brinkerhoff and Yeager 1993, 13). Thus, from the NEAP's inception biodiversity conservation was a top priority, although funding directed at biodiversity quickly exceeded expectations.

Building and Undermining Institutional Capacity

A core priority of EP1 was to construct a policy, regulatory, and institutional framework for a broad-based environmental program. The principal donors—including the World Bank, UNDP, UNESCO, USAID, French, German, Norwegian, and Swiss Aids, and the German Development Bank (KfW, formerly Kreditanstalt für Wiederaufbau), as well as conservation NGOs, WWF, and CI—provided financial and technical assistance to the DGEF; the Institut Géographique de Madagascar (Geographic Institute of Madagascar), the cartographic agency; and the Direction des Domaines (Land Titling Agency), which was responsible for establishing the boundaries of protected area and titling private property

(Brinkerhoff and Yeager 1993; Gezon 1997; World Bank 2003c). In addition, donors gave direct budget support to the government in exchange for the passage of a number of environmental laws. These included the Mise en Compatibilité des Investissements avec l'Environnement (MECIE) (Ensuring Investment Compatibility with the Environment) law of 1992, which required all national and foreign companies conducting activities with potential negative environmental impacts to undertake environmental impact assessments (Bekhechi and Mercier 2002), and the Forest Law of 1997, which consolidated forest legislation and established new permitting programs (Henkels 2001–2; Muttenzer 2002). The establishment of this legal and institutional framework comprised the foundation for the subsequent environmental programs.

However, the neoliberal push to reduce the state converged with donors' concern that the government was too corrupt to carry out its environmental policies, and donors leveraged the creation of several new organizations that they could fund directly and in which they could pay higher salaries than those of government employees (Falloux and Talbot 1993; Jolly 2015; Sarrasin 2007b). The World Bank's disbursement conditions specifically required the formation of three institutions. First, ONE took over from the unit that had integrated the NEAP preparations. It coordinated information about donor financing, served as a liaison among donors, NGOs, and government agencies, and distributed information about NEAP's activities to members (Falloux and Talbot 1993; Greve 1990; MDS 1990). It also oversaw environmental education and research programs. Second, based on a small pilot model conducted during the NEAP preparations, the Association Nationale d'Actions Environnementales (ANAE) (National Association for Environmental Actions) was set up to oversee agroforestry, soil conservation, watershed management, and small rural infrastructure projects paid for by a national environmental fund. Finally, ANGAP was established to superintend and coordinate the management of selected park reserves as well as international conservation and development projects (Brinkerhoff and Yeager 1993; Brinkerhoff 1996; Gezon 1997).

However, as EP1 actors sought to work around inefficient systems in the short term, they created long-term problems. The invention of these new organizations caused conflicts over budgets, staff, and transfers of responsibility; the introduction of new vested organizational interests; and a lack of interagency coordination (Andriamahefazafy and Méral 2004; Sarrasin 2007b). In particular, the formation of ONE led to tensions between it and the Ministry of Environment over NEAP coordination, which was reinforced by the fact that ONE was insufficiently staffed to manage the process (Brinkerhoff 1996; Falloux and Talbot 1993; Sarrasin 2007b). Likewise, the creation of ANGAP strained relations with the forest service. Reflecting back on the decision to create new organizations twenty-five years later as the illegal hardwood and gemstone trades were expanding, many interviewees felt it had been a mistake not to invest in the forest service, despite its shortcomings. A conservationist who had been involved in the NEAP's early negotiations explained the rationale for avoiding the forest service but also confessed that it had been a strategic error: "The creation of ANGAP in many ways originated from the lack of desire on the part of everyone to work with [the minister of livestock, water, and forests]. . . . But today, twenty-three years later, we're creating new categories [of parks] under [the Department of Water and Forests]. So they have a biodiversity agenda yet again."[5] Even a World Bank program evaluation stated that "the greatest risk to [the] development outcome of the entire Environmental Support Program is the lack of financial sustainability of the newly created environmental agencies" (World Bank 2007, 13).

Evaluations of the NEAP reiterated the need to combine biodiversity conservation efforts with rural development in order to address ultimate and proximate drivers of land use change. The early USAID consultants Brinkerhoff and Yeager (1993) argued that the NEAP was too focused on biodiversity and protected areas: "The Environmental Charter clearly incorporates the conservation-development policy linkages. However, the environmental action plan and the donor-funded projects being implemented to achieve its goals are weighted toward natural resource conservation policies, emphasizing 'green' issues and protected areas" (26). An evaluation of the EP1 echoed these concerns: "Over the past six years, the bulk of the effort under the National Environmental Action Plan has been focused on sectoral projects in the environment and natural resources sector and particularly on parks and protected areas. Yet there is growing awareness that the strategy for the environment has not paid enough attention to the underlying problems that cause environmental degradation" (Shaikh et al. 1995, 2).

With this goal in mind, parties to EP2 promoted the regionalization and decentralization of environmental management and rural development (Cowles et al. 2001; Medley 2004; Pollini 2011). A Madagascar government official recalled, "The idea during EP2 was to identify the origin of pressures [on forests] and accordingly to establish an action plan that would integrate development and environment."[6] Two key programs were initiated to implement regional planning: Appui à la Gestion Régionalisée de l'Environnement et à l'Approche Spatiale (AGERAS) (Support to Regional Environmental Management and the Landscape Approach) and Fonds Régional d'Appui à la Gestion de l'Environnement (FORAGE) (Regional Funds to Support Environmental Management). AGERAS established participatory planning committees comprised of communal and regional representatives to analyze environmental problems and to develop strategies to address these problems. FORAGE complemented AGERAS by providing a quick-dispersing grant fund to implement these actions (Andriamahefazafy and Méral 2004; Cowles et al. 2001). These regionally focused programs sought to integrate conservation and rural development in order to redress drivers of degradation, particularly forest loss, but they also led to a retrenchment of donors in regional zones during EP1, with the Americans in Fianarantsoa and Moramanga, Swiss in Menabe, and Germans in Vakinankaratra (Kull 2014; Moreau 2008).

The EP2 agenda was overwhelming. During the first few years of EP2 core donors funded fourteen programs.[7] In an effort to coordinate the donors' disparate activities, all donor assistance was to be channeled through seven Agences d'Exécution (AGEX) (Executing Agencies) (Gaylord and Razafindralambo 2005).[8] However, when donors funded executing agencies individually, coordination became more difficult (Hufty and Muttenzer 2002). Furthermore, programs not implemented through AGEX were not included administratively in coordinated planning efforts, including USAID's three principal environmental support programs: Miray (protected areas), PAGE (environmental policy), and Landscape Development Interventions (LDI) (BATS 2008). Finally, a massive exercise aimed at measuring the impact of the various environment projects created countless indicators of progress. Senior Madagascar government officials complained that there were so many indicators in EP2 that it was hard to tell what was most important.[9] Moreover, the focus on measureable short-term results discouraged investment in critical long-term institution building. A multidonor and government review in February 2001 concluded that, with its fourteen components and seven implementing agencies, EP2 was too ambitious. The overall program was downsized, and programs such as AGERAS and FORAGE, along with components of research and training, land tenure, marine conservation, environmental impact assessment, and monitoring programs, were eliminated (World Bank 2007).

This downsizing marked a critical turning point, one in which the environment agenda refocused on the four main programs that had funding: increasing rural capacity in soil and water management, improving communal and multiple use forest management, expanding the protected area system, and developing institutional environmental capacity at the national and regional levels (Gaylord and Razafindralambo 2005; World Bank 2003c). Then, during the transition from EP2 to EP3, donors decided to exclude previously supported rural development activities, and most of the seven EP2 executing agencies were eliminated or completely privatized. Some of AGERAS's activities were taken on by the Malagasy independent organization Service d'Appui à la Gestion de l'Environnement (SAGE) (Environment Management Support Service), particularly the oversight of CBNRM (Blanc-Pamard 2009), and ANAE, which had been responsible for soil and water conservation projects, was privatized and told to sell its services to potential buyers.[10] In contrast, donors continued to finance ANGAP as a "core agency," even though the eventual goal was for it to become independent from foreign funding and influence.

An evaluation of EP2 by the World Bank echoed the concerns that had been raised during EP1: "The conservation of natural resources cannot be planned in isolation. Coordination with other national programs, especially rural development, is critical" (World

Bank 2007, xii). Yet the graduation of a number of executing agencies resulted in the abandonment of agricultural intensification activities and a concentration on biodiversity conservation, pursued through the expansion of protected areas (Pollini 2007, 2011). Between the inception of EP1 and the Durban announcement that Madagascar would triple its protected areas, donors and the government invested US$75 million on expanding protected areas and an estimated average US$3 million per year went directly to ANGAP (Carret 2003). In EP3 the former president's initiative to triple protected areas took over much of the environmental agenda, and, again, the high-level politics superseded efforts to integrate conservation and development.[11]

Mainstreaming the Environment

As a result of the political violence in 2002 and the accompanying delays in processing donor funding, the third and final phase of the NEAP was postponed until 2003, and in the World Bank's case until 2004 (Méral 2012). On paper EP3 aspired to "mainstream the environment" more broadly into overall government policy. In line with the vision of environmental mainstreaming laid out in the environmental charter of 1990, no EP4 was planned. The Ministère de l'Environnement, Eaux et Forêts (MinEnvEF) (Ministry of Environment, Water, and Forests) was to take over management of the environmental agenda by the end of EP3, and donors were to contribute to an agenda determined by the Madagascar government rather than one determined by the joint government–donor committee.[12]

In EP3 the donor–government planning and coordination committee, or Comité Conjoint, was presided over by a representative of the MinEnvEF and a representative of the donors. The USAID environmental officer Lisa Gaylord held the position at the time of this research.[13] Once again donors supplanted ONE with another institution with which they could work directly: the Cellule de Coordination du PE3 (EP3 Coordination Cell), and ONE was assigned to manage environmental education programs, to ensure compliance with the MECIE, and to develop regional "state of the environment" reports (Andriamahefazafy and Méral 2004).[14]

The environmental agenda was reorganized again, this time around seven strategic objectives focusing on sustainable development, forest ecosystems and water resources, protected areas and conservation sites, coastal and marine ecosystems, environmental education, sustainable finance, and environmental policies and governance (Gaylord and Razafindralambo 2005). In contrast to EP2, all EP3 donor-funded activities, assuming they matched with elements of the program's logical framework, were to be regarded as part of NEAP irrespective of how they were channeled (BATS 2008). Picking up the emphasis on measuring results initiated during EP2, donors were to measure their accomplishments against standardized indicators, such as the rate of deforestation, degradation of critical habitats, quality of biodiversity in protected areas, households targeted for soil conservation, tourist revenues generated, and application of the environmental impact assessment law. Each of these general categories contained more specific outputs toward which programs were supposed to strive and which were summarized in an elaborate matrix of strategic objectives and outputs. Madagascar government agencies and foreign aid donors were supposed to contribute funds to these efforts. As in EP2, government and donor interviewees complained that developing and measuring the indicators in EP3 took an inordinate amount of time.[15] It also directed investment toward things that could be measured and away from long-term capacity building.

In EP3, 58 percent of donor funds were directed to forest management and protected areas (Tableau Récapitulatif des Contributions des Bailleurs de Fonds au PEIII 2003), and many of the other programs were indirectly related to the biodiversity agenda. In the transition from EP2 to EP3, research was eliminated as a core priority, a fact several Malagasy and foreign scientists emphasized in interviews: "There is no money that goes into the university. No one wants to pay their electricity there. Yet, people themselves have to be trained if you are going to have a functioning [environmental program]."[16] It seems ironic that a program pushed originally by scientists ended up underfunding Madagascar's universities. Madagascar government officials also expressed frustration at their inability to direct funds toward a broader range of the country's identified environmental priorities, especially reforestation and urban pollution. As one official complained, "Everywhere there are organizations who want to finance only biodiversity. We want to have funds, so we have to play the game."[17] A senior Madagascar government official in the MinEnvEF summarized the situation as follows: "In EP3, information, communication, education, urban environmental issues, urban pollution, and management are not really funded. Most of [the donors'] priority is biodiversity. They are not really interested in people's lives."[18] In 2005 and 2006 the government made a number of public appeals to existing donors to fund these priorities.

EP3 continued the regionalization effort by seeking to integrate environmental issues in newly formed regional government structures (Gaylord and Razafindralambo 2005),[19] although, again, with insufficient investment in regional state capacity building. Assisted by ONE, each region compiled a state of the environment report which summarized environmental information for each region, and each developed a Plan Régional de Développement (Regional Development Plan) designed to set regional priorities, including environmental ones.[20] However, even as the regions were tasked with these activities, state, regional, and local branches could not manage them without technical assistance from donor-funded contractors or NGOs or both. Despite fifteen years and millions of dollars of environmental foreign aid, many regional bureaucrats still worked without phones, faxes, and computers. The limited budgets of regional government contrasted sharply with those of international NGOs and contractors and parastatal organizations, which were, according to one World Bank report, "fifty times bigger" than that of the forest service (World Bank 2003a, 99). As Moreau (2008) captures beautifully, an upper-level manager in an international NGO earned a salary more than ten times the average salary of a civil servant: "WWF's offices hum and ring with the noise of computers and cellular telephones; while in an adjoining room agents for Waters and Forests use an old typewriter and carbon paper. They are forced to make do with an old motorbike for their rounds in a region and only earn a salary equivalent to that of a driver working for an NGO" (55). Even as donors spent millions on the environment program, the entire annual budget of MinEnvEF was only a few hundred thousand U.S. dollars (Lindemann 2004; World Bank 2004a).

This financial discrepancy reinforced centralized and donor control of the regionalization process even as the goal was to devolve authority. The result of such relative inequality in regional capacity was that although programs were aimed at modifying the behavior of rural peasants, management decisions about how to do that were made in Antananarivo, often by non-Madagascar state actors (Corson 2011; Hufty and Muttenzer 2002). Interviewees critiqued the "Antananarivo-centric" nature of the NEAP as a main obstacle to success in the field,[21] citing, "A huge disconnect—a cultural gulf—between the regions and Tana (the capital, Antananarivo)"[22] and "ignorance of regional differences."[23] They complained that the vast majority of funds remained in Antananarivo: a regional government official said, "The World Bank works with USAID; USAID works with CI; CI works with Miaro [a USAID conservation program]; and Miaro works with the operators. So you see how the money is spent. Just one thousandth of the money reaches the rural areas."[24] In this regard, even as these actors pushed to devolve management in the 1990s they centralized funding decision making. The strong national-level donor coordination created by the NEAP ultimately focused attention on meetings in the city rather than ensuring the participation of regional and rural actors in environmental planning processes.

Yet even in the capital environmental officials were often unable to influence overall government policy (Lindemann 2004). Each government ministry had an environment unit that was responsible for ensuring that ministry programs adhered to environmental laws, providing environmental management tools, and disseminating environmental information within the ministry. As of 2006 twenty-nine units had been established in sectoral ministries and the office of the prime minister, and plans to create a number of regional environmental cells were under way (Belvaux 2006; Mercier 2006). While the merging of the Ministry of Water and Forests and the Ministry of Environment in 2003 brought the forest service into an environmentally minded ministry, there was minimal coordination between the old General Directorate of Water and Forests and the new environmental portion of the ministry.[25] Finally, while a number of intersectoral policies had been drafted, environmental policies for the major sectors of agriculture, tourism, urban development, and forestry had not yet been implemented by the end of EP2 (Mercier 2006; World Bank 2007). By EP3 the government still lacked institutional and financial capacity, above all at the regional level, to carry the program forward without donor support, which was problematic as mining and timber exploitation escalated (Duffy 2005, 2007; Randriamalala and Liu 2010; Tegtmeyer et al. 2010).

Exploiting Madagascar's Natural Resources

Strong state incentives to attract foreign investors and minimal state infrastructure to enforce regulation have combined to make the sustainable management of the rapidly expanding extraction of Madagascar's natural resources one of the country's greatest upcoming environmental challenges. Structural adjustment reforms that advocated rapid economic growth through reduced state interference, an increase in foreign direct investment, and liberalized trade have simultaneously created incentives for natural resource exploitation and reduced the state's capacity to manage that exploitation (Duffy 2007; Hufty and Muttenzer 2002; Moreau 2008; Sarrasin 2006a). They have also increased the vulnerability of farmers producing export crops to price fluctuations in world markets (Feeley-Harnik 1995; Hufty and Muttenzer 2002; Sarrasin 2005). A range of economic policies have been implemented to attract investors, including reducing taxes, royalties, and other fees; eliminating restrictions on the repatriation of profits; and strengthening investor rights (Ferguson et al. 2014). While natural resource rents as a percentage of Gross Domestic Product (GDP) increased steadily from 3 percent of GDP in 2001 to 8.8 percent in 2012 (World Bank 2011b), the vast majority of wealth generated by Madagascar's natural resources still leaves the country (Ferguson et al. 2014).

Even as the mining sector has historically comprised a relatively small portion of Madagascar's GDP, large-scale mining is anticipated to play a major role in attracting FDI (Sarrasin 2009). Mining exports doubled between 1996 and 2000, and investments increased on average by US$3.6 million per year between 1990 and 2001 (Sarrasin 2009, 152). In 2009 exploratory and exploitation mining permits covered more than "three-quarters of the country" (Vega Media 2009), and "Between 2002 and 2009, the growth in awarded research permits alone corresponds to . . . over 35 percent of Madagascar's total surface area. Prior to the 2008 crisis, almost the entire mining surface in the country was thus allocated to permit holders" (World Bank 2010, 40). Many of these permits are in protected areas (see chapter 7). While the increase in demand has led to revenue of "an estimated US$5 million in 2008 and 2009" (World Bank 2010, 41), the mining code requires companies to pay only 2 percent of the price of mined product in tax, 1.4 percent going to regional government and 0.6 percent to the state, with the regional portion further distributed to commune, region, and province (Ferguson et al. 2014).

The mining industry in Madagascar is comprised of two relatively distinct sectors: large-scale mining, which focuses primarily on various industrial ores,[26] and a small-scale, informal, unregulated gemstone-mining sector. Large-scale industrial ore mining has generated the majority of revenue, while a very small percentage has come from small-scale mining, as most gemstones are exported via illegal traffickers and/or exported as rough stones and traded to gem dealers. Two large foreign mining projects have dominated the large-scale mining sector: the QIT Madagascar Minerals (QMM) (an affiliate of the multinational Rio Tinto Zinc [RTZ Corporation]) project to extract mineral sands for titanium and ilmenite in Madagascar's southeastern coast; and the Sherritt (formerly Phelps-Dodge/Dynatec) mine for cobalt and nickel in Ambatovy in eastern Madagascar. In addition, oil and gas exploration has expanded rapidly: the Madagascar Oil tar sands project claims to be holding more than a billion barrels of oil (Webb 2010). Drilling for coal, uranium, oil, gas, and iron; production of chromite and cement; and oil sands for bitumen and offshore drilling are all anticipated to increase in the coming decade (Blanc-Pamard 2009; Canavesio 2014; Duffy 2007; Sarrasin 2003, 2006a; Yager 2009).

The gemstone mining sector employs hundreds of thousands of people. Madagascar is the world's top-ranked sapphire producer, supplying roughly half of the world's sapphires. It is also a source of a number of other precious and semiprecious stones (Canavesio 2014; Sarrasin 2003, 2006a; Yager 2009).[27] Most of the gemstone market is illegal: US$100 million worth of precious stones were smuggled out of Madagascar in 1999 alone; roughly US$4 million worth of stones changed hands each day in 2001 when trading in Ilakaka was at its height; and by 2004 gem mining covered 4,000 square kilometers (Duffy 2005, 2007). While legal exports doubled between 1996 and 2000, from US$16 million to US$37 million, estimates from illegal exports reach US$200 to US$500 million annually (World Bank 2003b).

Reforms to the legal and institutional framework for the mining sector began in 1998 and resulted first in the Mining Code of 1999 (Repoblikan'i Madagasikara 2005d, e). Over the past fifteen years the World Bank has pushed additional reforms in order to promote exports, encourage private sector development, disseminate royalties to the regions, and improve governance and transparency, particularly in small-scale and artisanal mining. Its programs have leveraged revisions to the Mining Code and the creation of the Madagascar Mining Registry Office and the Madagascar Gemology Institute, linked to the Gemological Institute of America, in order to train an estimated four hundred gem cutters a year (Canavesio 2014; Duffy 2007; Ferguson et al. 2014; Sarrasin 2006b, 2007a, 2009; Vega Media 2009). Environmental regulations have concentrated on minimizing and making reparations for damage, in protected areas expressly, through environmental impact assessments (Sarrasin 2006b).

However, "after more than 20 years of structural adjustment, the governments' institutional capacity has been impaired.... [A]lthough Madagascar has strict environmental protection legislation, its implementation is far from certain, especially in the context of accelerated liberalization such as that entrenched in the mining code adopted in 1999" (Sarrasin 2006a, 395). In addition, the political regime change in 2009 increased opportunities for illegal resource extraction (Waeber 2009). As the World Bank reports, "Since 2006, the transparent mining right management has not been regularly applied" and "following the 2009 political crisis, the permit management process appears to have reverted to a more discretionary management" (World Bank 2010, 40–41).

Like illegal mining, illegal forest trade, particularly of rosewood, palisander, and ebony in northern Madagascar, has been a known issue for decades, one complicated by high-level political corruption, threats to those trying to expose it, and the alternating of state proclamations against illegal timbering and state orders authorizing exceptional exports (Schuurman and Lowry 2009; Zoo Zürich 2009b). Estimates of the percentage of timber extraction that is illegal have ranged from 60 percent (World Bank 2003a, 99) to 90 percent (Becker, Myers, and Pierson 2005, 7). Most of it has been destined for China, at one point for the 2008 Olympic Games but also for craft furniture and musical instruments (Débois 2009; Innes 2010; Schuurman and Lowry 2009; Wilmé et al. 2009; Zoo Zürich 2009a). Much of the illegal forestry has taken place in protected areas, and it increased with the political turmoil after Ravalomanana was ousted from power (Freudenberger 2008; Global Witness and EIA 2009; Randriamalala and Liu 2010; Schuurman and Lowry 2009) (figs. 5.1, 5.2).

Donors, including USAID, have leveraged various reforms in the forest service, such as improvements in regulations, issuing procedures, enforcement, and transparency (Hufty and Muttenzer 2002; World Bank 2004b), and they have funded, albeit insuffi-

Figure 5.1. Illegal logging in the Fandriana–Vondrozo Corridor. Photograph by Vokatry ny Ala.

*Figure 5.2. Boards waiting for pickup in the Fandriana–Vondrozo Corridor.
Photograph by Catherine Corson.*

ciently, the countrywide forest land use zoning (BATS 2008; Hufty and Muttenzer 2002). The IMF included conditionalities for forest management reforms to improve transparency, accountability, and governance in Madagascar's Highly Indebted Poor Country (HIPC) debt initiative (Winterbottom 2001). However, as Randriamalala and Liu (2010) underscore, the reduced state, combined with low government salaries, has meant that the state has little capacity or incentive to enforce regulations: "Under the influence of environmental donors (who must demonstrate visible results during the lifetime of their projects), the government enacts sound forestry legislation and takes restrictive regulatory measures, but it does this without having the capacity, human OR moral, to apply them" (21).

Regional and local forest service offices have faced severe shortages of staff, equipment, training, and funds, and insufficient salaries have reinforced corruption.[28] As a USAID contractor report summarizes, "The current system suffers from many problems, including a significant shortage of human, material, and financial resources, outdated regulations, an insufficient set of incentives for forestry agents, limited communication, non-functional field-level control of the harvesting and movement of forest products (chain of custody), weak judicial pursuit of offenders, unsupported partners, insufficient training, and virtually no regular implementation of enforcement activities. In this operating environment, there have been few to no disincentives for illegal harvesting of forest resources, and there has not been a level playing field for loggers who do want to respect existing regulation" (Becker, Myers, and Pierson 2005, 4). Additional issues with the legal permitting process have ranged from their three- to five-year duration, which encourage exploiters to take timber out quickly, to the lack of boundary marking before and limited oversight during operations, to the networks of tracks left that facilitate access to the interior after operations (Bertrand et al. 2014; World Bank 2003a).

As former president Ravalomanana's initiative to triple protected areas began to monopolize the environmental agenda, it superseded initiatives such as the forest sector reform and forest zoning process that were designed to redress some of these issues.[29] This was evident in a meeting of the Comité Conjoint in 2005, at which the venue was filled for a discussion of SAPM but then emptied during the discussion of the forest sector reform. Funded in part by USAID through the Jariala project, the forest sector reform aimed at zoning forestlands for various uses, including conservation, commercial forestry reform, and protection of future fuelwood supplies.[30] Regional DGEF agents and consultants expressed frustration that they had hoped to complete the forest zoning process before SAPM, but owing to political pressure to implement SAPM quickly they had to prioritize delimitating the new protected areas rather than looking at long-term use planning or managing the forest for multiple uses.[31]

In this manner high-level political attention undermined the ability of lower-level program managers to build sustainable management programs. A representative of a Malagasy conservation NGO who had worked on Madagascar conservation issues for many years reflected on this imposition of SAPM before the forest zoning process was complete: "The conservation community has a tendency to jump onto a new thing suddenly. . . . Now most people are focused on SAPM, and the reform of the forest service has dropped. . . . Yet the forest service reform is the basis for SAPM. . . . Similarly, there are good lessons to be learned from the community management transfers, yet no one is working on them. All attention is on the Durban Vision."[32] Similarly, the former head of the French Service de Coopération et d'Action Culturelle (French Cooperation and Cultural Action) in Madagascar wrote, "A recent assessment mission found that the EP3 experienced 'a modest and unbalanced progress' because it was too focused on the implementation of the president's agreement to expand the protected area network at the expense of a significant improvement in environmental and forest governance" (Belvaux 2006, 5, author's translation). As I explore further in subsequent chapters, the establishing of new parks that lack effective management capacity will do little to stem either legal or illegal mining and forestry. Regulated and illegal mining and forestry have all taken place in protected areas both beyond the control of the government and in collusion with government officials (Duffy 2005; Global Witness and EIA 2009; Walsh 2005).

In 2009 Global Witness and the Environmental Investigation Agency (EIA) conducted an undercover investigation into the trafficking of rosewood, palisander, and ebony, producing video evidence and testimony from local communities and naming a series of people, including well-known local politicians and vanilla and lychee traders involved in illegal trading (Global Witness and EIA 2009). Donors, scientists, and conservation organizations then issued a series of statements calling on the government to stop illegal forestry (e.g., CAS et al. 2009; Global Witness and EIA 2009; Wilmé and Waeber 2010; WWF et al. 2009; Zoo Zürich 2009a). The report's authors also helped to convince members of the U.S. Congress to pass a nonbinding resolution that condemned the illegal extraction of Madagascar's natural resources, requested importing countries to intensify their inspection and monitoring process to ensure they do not import wood from Madagascar, and urged the public to boycott products made of Malagasy wood (U.S. House of Representatives 2009). In 2013 the members of the Convention on International Trade in Endangered Species agreed to list Madagascar's populations of rosewood, palisander, and ebony in appendix-II, which restricts their trade but does not require permits to import them (fig. 5.3). While donors have indicated a recent willingness to tackle the issue of illegal extraction, the need to incorporate sustainability into economic planning remains.

Figure 5.3. Toamasina Port, from which illegal timber is shipped. Photograph by Desmond Fitz-Gibbon.

Isolating and Reembedding the Environment in the Market

The conservation community's concentration on protected areas has isolated the environment from economic development planning and then reembedded it in market exchange via green commodities, such as carbon credits and biodiversity offsets. This strategy has carefully avoided the core issues of sustainable use and growth and even sanctioned exploitation.

Throughout the NEAP's three phases a number of actors have underscored the need to evaluate the impact of fiscal systems, economic incentives, and subsidies on Madagascar's environment. At the EP1 midterm review in 1993, USAID Director George Carner pointed to the need to redress weak connections between the first phase of the environmental program and Madagascar's overall macroeconomic planning (Greve 1994). The staff appraisal report for the World Bank's initial NEAP funding similarly expressed concerns about the potential negative effect of liberalized trade on Madagascar's environment: "It is still too early to say whether the incentive mix engendered by more exposure to market forces will continue to encourage the destructive process, or whether it will help slow it down" (World Bank 1990a, 103). Finally, a consultant report from EP3 admitted that "the government's push to increase foreign direct investment through mining—and soon, more overt petroleum exploration—is a huge challenge in the face of maintaining and protecting Madagascar's unique natural environment. The MAP [Madagascar Action Plan] lays out ambitious targets, that if not managed properly, could have disastrous consequences" (BATS 2008, 126).[33]

Despite these warnings, across the NEAP's three phases there has been limited discussion of ensuring that environmental sustainability is incorporated into macroeconomic planning. The MECIE has been the primary tool invoked to ensure sustainable economic development, but the state has had minimal incentives and ability to enforce regulatory legislation in the face of the rapidly growing natural resource industries. Even as the NEAP was touted as a model program, its advocates paid little attention to environmental economic tools like green national accounting or green GDP. Instead, overarching Madagascar government policy, as seen in visions of the Poverty Reduction Strategy Paper (PRSP), MAP, and Madagascar Naturellement! (Madagascar Naturally!), represented politically successful strategies that aligned key political players from across the government and donor community around a narrative that simultaneously blamed the poor, appealed to conservation donors, and enabled the pursuit of resource-based growth and liberalized trade in line with IFI policies.

In 1996 the World Bank and IMF launched the HIPC initiative in an effort to address unsustainable debt in the world's most heavily indebted and poorest countries. Beginning in 1999 the World Bank and IMF began requiring grant and loan recipient countries to develop PRSPs that described how their economic growth strategies would also reduce poverty (Christiansen and Hovland 2003; Craig and Porter 2003; Froger and Andriamahefazafy 2003; Porter and Craig 2004). PRSPs eventually became major planning documents for multilateral and bilateral aid as well as aid-recipient government policy. In line with the HIPC enhancement program of 1999, an enhanced debt relief package was issued for Madagascar in

2000. It was subject to certain conditions, including continuation of structural adjustment to promote rapid GDP growth and increased FDI as well as the completion of a PRSP that included reforms to promote transparency in the governance of mining, fishing, and forestry sectors. The Madagascar government submitted its initial "Interim-PRSP" in November 2000 and its full PRSP in September 2002. As in the case of the NEAP, the preparations entailed a participatory process involving bureaucrats, politicians, private sector representatives, NGOs, and others (Froger and Andriamahefazafy 2003; Republic of Madagascar 2000, 2003).

The PRSP was based on the premise that "export growth contributes to reducing poverty while at the same time protecting biodiversity" (Sarrasin 2009, 154). It emphasized the importance of integrating environment into regional planning, and it argued that sustainable management of natural resources was fundamental to Madagascar's economy: "At least 50 percent of income in the Malagasy economy is directly dependent on natural resources. If one considers the labor market, 9 jobs out of 10 are in the sectors that depend directly on natural resources. Eradicating poverty in Madagascar will necessarily be contingent on the sustainable management of this capital that is key to the national economy, namely the environment" (Republic of Madagascar 2003, 97). However, even as it advocated governance reform it did not specify how to ensure that the projected 8 percent growth rate would be achieved through sustainable use of natural resources or how the environmental dimension would be effectively integrated into other sectors, such as mining, energy, and water (Lindemann 2004). Instead, it emphasized protecting natural resources from unsustainable rural community practices like shifting cultivation by establishing protected areas (Republic of Madagascar 2003).

Like the PRSP, Ravalomanana's framework Madagascar Naturellement! presented Madagascar as a site for simultaneous conservation and resource extraction. Announced at a conference in France in January 2005, Madagascar Naturally! laid out a plan to market the country to potential investors as a "natural" site with "natural" riches, and it proposed to drive economic growth by increased production and quality of "natural products" and via resource-based industries, such as food and nonfood agroindustrial production, nonprecious and precious stones, textile industries, and ecotourism. Again, the overall program emphasized growth driven by natural resources, while its environmental section focused on protected area expansion.

The Madagascar Naturally! vision was then incorporated into the MAP, a government agenda which established the government's direction and priorities from 2007 until 2012 (Repoblikan'i Madagasikara 2006a). The MAP concentrated on eight commitments, one of which was environment. Like the NEAP, it emphasized the interests of donors and NGOs, and its stated environmental goals were as follows: (1) increase protected areas to conserve land, lake, marine, and coastal biodiversity; (2) reduce degradation of forests, wetlands, lakes, marine, and coastal resources; (3) mainstream the environment into all sectoral plans; and (4) strengthen the effectiveness of forest management (Repoblikan'i Madagasikara 2006a). It articulated a vision that linked biodiversity conservation and natural-resource-driven growth in a partnership: "Madagascar will be a world leader in the development and implementation of environmental best-practice.... The world looks to us to manage our biodiversity wisely and responsibly and we will.... Given the Government's vision—Madagascar Naturally!—we will develop industries around the environment, such as eco-tourism, agri-business, sustainable farming practices, and industries based on organic and natural products. These industries and activities will minimize biodiversity damage and maximize benefits for the nation and the people" (Repoblikan'i Madagasikara 2006a, 97). Again, it offered no details on how these industries would minimize their impact on biodiversity as they expanded or on how the markets for these products would be developed.

As conservationists endeavor to find economically viable alternatives to resource extraction that demonstrate the economic value of biodiversity and to raise conservation funds, they have embraced new market mechanisms and financial instruments. Ecotourism and bioprospecting have been well-established ways to extract wealth from conservation sites in Madagascar (Neimark 2010, 2012; Sarrasin 2005). Emergent approaches to securing conservation funding include payment for ecosystem services, REDD+, and biodiversity offsets (Brimont and Bidaud 2014; Ferguson 2009; Ganzhorn 2011). In 2001 USAID financed the first feasibility studies for a project to fund conservation through carbon credit sales in the Makira protected area. The Makira Carbon Company was launched in 2008, and the project received accreditation through the Climate, Community and Biodiversity Alliance standards to sell on the voluntary carbon markets. There are four other REDD+ pilot projects in Madagascar (Brimont and Bidaud 2014; Ferguson 2009; Méral et al. 2009). At the same time, QMM has initiated a biodiversity offset program that seeks to compensate for the environmental damage of its mine by creating new protected areas that offset its ecological impact (Kraemer 2012; Seagle 2012; Waeber 2012). Finally, in 2005 WWF, CI, and the Madagascar government established a Madagascar Biodiversity Fund, which, with help from the World Bank, France, and Germany, had raised US$50 million by the end of 2014, the funds to be managed by JPMorgan Chase and Co. and the interest on investments to be available to finance conservation (Méral et al. 2009; Méral 2012).

However, as these new approaches endeavor to raise funds for conservation, particularly in areas where the commercial activities of peasants are simultaneously being restricted, they introduce issues around equitable distribution of wealth, both within communities and along the commodity chain (Bertrand, Horning, and Montagne 2009; Brimont and Bidaud 2014). Even when rural communities may be the intended beneficiaries of market-based conservation, any projected benefits are subject to price fluctuations in global markets and may not generate sufficient revenue to compensate rural people for their loss of access to land and resources. Furthermore, as Scales (2014b) argues in relation to ecotourism projects, even when benefits accrue to communities, the costs of the loss of access to resources are usually felt by households. Such projects also empower conservation organizations as mediators between project sites and international finance networks (Kraemer 2012; Seagle 2012; Waeber 2012). Finally, they secure the bulk of profits that might be generated from such projects for investors in the global North (Wang and Corson 2014). They enable a variety of external actors—from pharmaceutical companies to financial traders to transnational conservation NGOs—to secure wealth from Madagascar's natural resources and, as they do so, use the moral justification of conservation to restrict peasants' rights to profit from resources. Ultimately, they do not displace but exist in parallel to both the regulated and illegal resource extraction that takes place in Madagascar's protected areas.

Neoliberal Environmentalism

While there are numerous success stories associated with the NEAP, including the establishment of a legislative and institutional framework for environmental management, the program's primary achievement was how it brought together and aligned donors, government, NGOs, consultants, scientists, and others around a cohesive environmental agenda. Through meetings, negotiations, reviews, and other events, these actors not only negotiated programmatic priorities but also developed political alliances and friendships that sustained long-term political attention to the plight of Madagascar's environment.

Yet the need to maintain these alliances undermined the NEAP's ability to tackle critical environmental issues. As donors, governments, and NGOs placed both the blame for environmental damage and the responsibility for environmental repair on rural peasants, they reinforced a legacy of precolonial, colonial, and postcolonial policies that limited rural peasants' activities while facilitating large-scale commercial extraction. Concurrently, by eliding the environmental impacts of commercial exploitation, the PRSP, Madagascar Naturally!, and MAP engaged key public, private, and nonprofit actors, many of whom would otherwise have been alienated by a commitment to stronger government regulation of resource exploitation. In this manner, the emphasis on conservation of biodiversity as being accomplished through the expansion of protected areas became a way to define "the environment" such that public–private and nonprofit entities alike could embrace it without confronting large-scale legal or illegal extraction.

The NEAP's creation and implementation in a period when the government was undergoing structural adjustment reforms fundamentally shaped the environmental narratives, priorities, strategies, and alliances that emerged in it. Numerous actors emphasized the need to accompany biodiversity conservation with rural development initiatives, to build government capacity to secure sustainable resource extraction, and to ensure that sustainability was a central concept in macroeconomic policy. Yet pressure to liberalize the economy, attract FDI, and reduce investment in state planning, management, and regulatory capacity left the government unable and unwilling to redress unsustainable (legal and illegal) natural resource extraction, including in protected areas. Instead, this approach isolated the environment in protected areas, giving external participants—from mining companies to illegal forest exploiters to carbon traders—a green stamp of approval to continue extraction.

Dependent on donors simultaneously pushing neoliberal reforms and biodiversity conservation, the Madagascar government carved a narrow political pathway for itself in government policies such as the NEAP, MAP, PRSP, and Madagascar Naturally! It avoided the issue of sustainable resource management in economic planning, opting instead to separate out the environment, as biodiversity and green products, while promoting rapid economic growth based on FDI and liberalized trade of natural resources. Ultimately, the NEAP and national development policies became means for restructuring political relations so as to enable rather than impede the rapidly growing extraction of its natural resources. The result was that, despite twenty years' effort and millions of dollars in investment, donors, governments, and NGOs alike were unprepared for the surge in natural resource exploitation that took place around the political crisis of 2009. Had the donor community prioritized building rather than reducing the state's capacity to manage the natural resource extraction sector in the 1990s, both state and nonstate actors might have been better equipped to redress the rush to exploit resources and the resulting vacuum in foreign aid that happened in relation to the political instability of 2008–9.

CHAPTER SIX

Creating the Transnational Conservation Enterprise

Environmental programs in Madagascar were spared only because of the Congressional biodiversity earmark. The earmark has been instrumental in assuring continued funding for the environment but has at the same time reinforced a relatively narrow biodiversity focus. In the absence of other funds, the Madagascar program has faced consistent difficulties in addressing complementary issues such as agriculture and economic growth. While transformation of Madagascar's economy might well have been impossible even with more robust agricultural and economic development funding, there can be no doubt that success on the environment front has been constrained by broader economic development failure, particularly in Madagascar's rural areas.

—KAREN FREUDENBERGER, *Paradise Lost? Lessons from Twenty-Five Years of USAID Environment Programs in Madagascar,* Executive Summary

USAID HAS BEEN A pillar of the Madagascar environment community since the mid-1980s. Numerous people working in Madagascar environmental politics have underscored that the financial and political backing of USAID and U.S.-based conservation NGOs has been instrumental to the NEAP's success.[1] A senior USAID contractor referred to USAID as "a pioneer of the environment program in Madagascar,"[2] and a recent consultant report called the agency "a primary champion and recognized leader of the environmental community in the country" (BATS 2008, 102). USAID financed the first NEAP donor-coordinating group, and, as the largest bilateral donor to the NEAP over its lifespan, it contributed around US$120 million (Freudenberger 2010; World Bank 2007), or about 32 percent of all donor funding between 1990 and 2003. When combined with the World Bank contributions, the total equated to about half of all donor funding of Madagascar's environment program (Andriamahefazafy and Méral 2004; Kull 2014). Its projected EP3 funding alone comprised 57 percent of all bilateral donors' contributions (Horning 2008a). USAID was also an intellectual leader. During EP3, USAID cochaired the NEAP's Comité Conjoint, or joint donor-government committee, and U.S.-based conservation NGOs and USAID contractors were also actively involved in shaping the NEAP's environmental priorities. The influence of the United States grew during Ravalomanana's presidency, as he sought partnerships with Anglophone organizations in education, the religious community, and donor groups in order to demonstrate a break with the country's French colonial history (Moreau 2008).

Most of USAID's investments focused directly or indirectly on biodiversity. USAID was a particularly important supporter of protected areas. It contributed 68 percent of funds for protected areas during EP1 (Andriamahefazafy and Méral 2004) and was a core funder of the protected areas expansion. U.S. conservation NGOs and government officials served on the Malagasy delegation that formulated the former president's WPC announcement, and it was the primary bilateral donor for SAPM, through a grant to CI that included subgrants to WWF-U.S. and WCS as well as to ANGAP (now the Malagasy Parks Board). Finally, USAID cochaired the technical committee for the SAPM.

Almost 100 percent of the USAID–Madagascar environmental program was funded by the biodiversity earmark over its lifespan (USAID 2008), which both protected and limited the agency's investments in Madagascar, as outlined in the above epigraph (see also chapter 4). The USAID mission opened in Madagascar only after the government there agreed to structural adjustment and specifically to liberalize its economy. The overall goal of the mission was to promote neoliberal economic growth in Madagascar. In the face of cuts to the mission's budget and personnel, the biodiversity earmark ensured that conservation remained central to the mission's portfolio. Nevertheless, as discussed in previous chapters, the strict guidelines on using these funds dissuaded investment in more comprehensive environmental approaches.

Representatives of U.S.-based organizations constituted a formidable intellectual and financial power behind the conservation program in Madagascar. Many of the scientists who catalyzed global attention to the plight of Madagascar's flora and fauna in the 1970s and 1980s were American, and many of them continue to influence Madagascar's environmental politics today. Building on the close ties forged by scientists and policy makers in the 1970s and 1980s, U.S.-based conservation NGOs and the USAID mission developed a mutually beneficial relationship that evolved through progressive conservation trends. USAID brought the NGOs into the world of foreign assistance across the globe by channeling its growing biodiversity portfolio through them (see chapter 4). In Madagascar, as the agency began involving NGOs in the design and implementation of its aid programs, it facilitated their rising influence in Madagascar's conservation politics, where they became influential donors themselves as well as members of the NEAP and SAPM coordinating committees. In Washington, as NGOs became more influential on Capitol Hill, they garnered support for the biodiversity earmark that funded the mission's environmental programs as well as for the USAID Madagascar mission itself.

My goal in this chapter is to show—by moving between policy-making sites in the United States and Madagascar—how the evolution of America's environmental investments in Madagascar was both constitutive and reflective of power relations between public and nonprofit organizations that transcended national boundaries.

I begin by discussing the history of early U.S. aid to Madagascar. I then analyze the progression of contributions by the United States to the NEAP, and I document how the relationship among NGOs, USAID, and the U.S. Congress evolved through various conservation trends. I evaluate impediments to the USAID mission's efforts to promote integrated environment and agriculture programs and the ways in which actors creatively worked around these constraints. I conclude by exploring the relationship between conservation politics in Washington, D.C., and in Madagascar.

The Inception of USAID's Environmental Program in Madagascar

In 1961, the year of its creation, USAID signed its first bilateral aid agreement with the Madagascar government. Its initial investments in Madagascar focused primarily on agriculture and food aid: between 1962 and 1978 it provided about US$20 million in infrastructure loans and P.L. 480 Title II food aid through the NGO Catholic Relief Services.[3] In 1972, however, USAID closed its Madagascar office in response to the second revolution, thereafter giving only food aid and making capital project loans out of its East Africa regional office in Nairobi, Kenya (IDCA 1981, 1985; U.S. Department of State 1979; USAID 1983a). After the government of Madagascar signed its first structural adjustment agreement with the IMF and satisfied USAID that the country was undertaking neoliberal economic reforms, the agency reopened a one-person office in 1984, managed by USAID Foreign Service Officer Sam Rea. Rea remembers that "Washington had been alarmed by the anti-Western stance of the Ratsiraka government. Our concern had particularly to do with Madagascar's strategic location on the major shipping lane through the Mozambique Channel. So as soon as Ratsiraka declared that he was willing to deal again with the West and consider economic reform, we had a strong incentive to offer renewed short-term help."[4] The agency began investing again in improving rice production and implementing a P.L. 480 Title I commodity program for rice (USAID 1984a, b).

Additional U.S. funds to Madagascar were conditional on the country's continuation of reforms. As Rea describes his assignment,

"First, I was to direct a program of short-term help and exploration; and second ... I was to prepare a program of long-term assistance, which AID could implement at such time as the U.S. was convinced that the Government of Madagascar was truly committed to policies of economic liberalization and reform. If this were to happen, AID would then be prepared to move ahead quickly with long-term development assistance at higher funding levels."[5] In 1987, citing the Madagascar government's decision to adhere to IFI reforms, the U.S. government declared Madagascar a "potential trading partner" and opened a full office, giving the Madagascar government balance-of-payment support for policy reform in agricultural markets and price liberalization (USAID 1984b, 1988b, 1992a, 1995b). Its first environmental program, implemented between 1984 and 1994, used slightly over US$1 million of P.L. 480 counterpart funds to finance five one-year projects that established nurseries and forest plantations as well as watershed and forest reserve management (USAID 1994a). USAID-Madagascar strategies published before 1986 focused primarily on agricultural productivity, infrastructure, and economic growth (USAID 1984a, b), but by early 1987 it was using counterpart funds to finance the local costs of the NEAP process (Rea 1998). By the time of the NEAP launch in the early 1990s the USAID-Madagascar mission had more than a dozen direct hire employees and programs in agriculture / economic growth, environment, health, and democracy / governance. The Peace Corps began operating in Madagascar in 1993 (Freudenberger 2010).

USAID's initial proposals for increased environmental investments underscored the need to improve agricultural productivity in order to combat the problem of shifting cultivation, which it identified as the primary threat to forests and biodiversity. In 1987 a proposal for an expanded "natural resource management and biological diversity" program argued, "We believe that the interactions between the agricultural sector and the component sectors of forestry, energy, and natural resources need to be addressed if Madagascar is to achieve sustainable development and the conservation of tropical forests and biological diversity" (USAID 1987, 56). It suggested three main areas of focus: "Improved agricultural productivity through intensive farming techniques, particularly rice production; the promotion of sustainable agricultural productivity among subsistence farmers through crop selection, soil conservation, and agroforestry activities; and the conservation of tropical forests and biological diversity" (USAID 1987, 57). Improving the ability of rural peasants to sustainably use resources became a guiding principle for the subsequent environmental program.

From Information Brokers to Political Advocates

Until the mid-1980s conservation NGOs acted primarily as political advocates, bringing attention to the plight of Madagascar's flora and fauna through studies, reports, and the media. The initial proposal for an expanded environmental program (see above) cited as rationales for the program two pieces of research by U.S.-based NGOs: a WWF-U.S. report of 1986 which asserted, "Human activities have already caused the extinction of 33 percent of Madagascar's primate fauna"; and a WRI report in 1985 entitled "Tropical Forests: A Call for Action," which recommended spending US$25 million over five years to preserve Madagascar's eastern rainforests. It also referenced the conference that WWF-International had sponsored in 1985 (see chapter 3) (USAID 1987, 53–55).

Yet the mid-1980s, as discussed in earlier chapters, was a critical historical moment in both the United States and Madagascar, a moment in which relations between the state and nonprofits began shifting vis-à-vis the biodiversity conservation agenda. In Madagascar preparations for the NEAP started soon after the 1985 conference, and the World Bank enlisted scientists, consultants, NGO representatives, and other nonstate actors in them. In the United States, Congress first authorized funds for biodiversity and forests in 1985, requiring that they be channeled through NGOs, and then, in 1986, it directed the agency to spend US$4 million on biodiversity and forests (U.S. Congress 1986a, b). The USAID mission's budget request for environmental funds in 1989 emphasized the importance of using NGOs to implement the environmental program. It stated, "[USAID-Madagascar] believes that PVOs are one of the most effective mechanisms to support natural resource activities. . . . Collaborative efforts on their part related to the conservation of tropical forests and biological diversity in the context of rural development would be very advantageous. These organizations have expressed interest in expanding their project portfolios into the natural resource / agroforestry sub-sector" (USAID 1987, 56). As USAID began responding to congressional directives, conservation organizations evolved from advocates and advisors to financial conduits and program implementers.

The collaboration between USAID and the conservation NGOs began with ICDPs, which emphasized development activities in park peripheries as incentives for people living in these zones to conserve park resources (Alpert 1996; Richard and O'Connor 1997; Swanson 1997). USAID became a leading sponsor of ICDPs across the globe, and in Madagascar the agency turned to the few U.S.-based conservation and development NGOs that had continued operating on the island during the political turmoil of the seventies, namely, WWF, Catholic Relief Services, and the Missouri Botanical Garden (MBG) as well as universities, including Duke, North Carolina State, Yale, and the University of Washington, which had been cataloguing and studying various species (Brinkerhoff and Yeager 1993; USAID 1987, 1994a).[6] In FY 1987 USAID provided US$400,000 to WWF-U.S. and the University of Madagascar's School of Agronomy for the ICDP at the Bezà Mahafaly Special Reserve, where Alison Richard had been managing a project in conjunction with the University of Madagascar since 1975 (Richard and Ratsirarson 2013).[7] The agency also funded a related ICDP at Andohahela managed by Yale University and the University of Washington (Fenn 2003; Kull 1996; USAID 1987, 1988b, 1994a). In FY 1988 USAID increased its funding, providing grants to Catholic Relief Services for an agroforestry project in west-central Madagascar; WWF-International and Catholic Relief Services for an ICDP project at Amber Mountain; MBG and the Lutheran church's agricultural development branch, called SAFAFI, for an ICDP on the Masoala peninsula; and Duke University in FY 1989 and, later, the National Center of Applied Research and Rural Development, usually referred to as FOFIFA, and North Carolina State for Ranomafana Park (Gezon 1997; Kull 1996; USAID 1987, 1988a) (fig. 6.1).

In addition to these ICDPs, CI, WWF-U.S., and USAID facilitated a number of debt-for-nature swaps, financial transactions in

Figure 6.1. Ranomafana National Park. Photograph by Desmond Fitz-Gibbon.

which a portion of a nation's foreign debt is forgiven in exchange for investments in conservation. From 1989 to 1996 nine commercial debt-for-nature swaps in Madagascar generated US$11.7 million in conservation funds. Under the first swap in 1989, WWF-U.S. and the Central Bank of Madagascar agreed to exchange US$3 million of Madagascar debt to fund four hundred local nature protection agents as well as additional forest service agents (Freudenberger 2010; Lewis 1999; McConnell and Kull 2014; Paddack and Moye 2003). The swap also granted funds to Bezà Mahafaly between 1994 and 2001 (Richard and Ratsirarson 2013). In 1990 CI conducted the second swap, and others followed soon thereafter (Feeley-Harnik 1995; Tucker 1994).

While USAID provided funding to the conservation NGOs, it also gave them political legitimacy. A former WWF official said, "We received a lot of support from USAID. This is how we got traction with the [Madagascar] government. Without the support of USAID, not just financial support, but also the interest that

USAID showed in the WWF program, I don't think we would be where we are today. . . . The government was not receptive to [conservation]; the international community was not entirely receptive to [conservation]; we just had a few countries supporting us . . . [and] by far the most important was the U.S. government with USAID."[8]

The introduction of the NEAP catalyzed a rapid increase in USAID environmental investments in Madagascar. In contrast to the one-hundred-thousand-dollar grants USAID awarded in the late 1980s, it awarded US$86 million dollars for the first phase of EP1 (USAID 1997d). As USAID scaled up funding, it began channeling millions of dollars through U.S.-based NGOs operating in Madagascar, contributing to their rising influence in Madagascar politics (Gezon 2000). A scientist who had worked in Madagascar for many years underscored the competition it created: "When Madagascar was put on the map, there was increasing money available, and with the increasing money came competition among the conservation organizations, like WWF, WCS, and CI. . . . They have become brokers of money, and USAID is responsible for the shift."[9] As "brokers of money," they developed a vested interest in shaping USAID's priorities, and their influence was reinforced by the declaration, in the mid-1990s, of Madagascar as a focal country of the USAID New Partnership Initiative, Vice President Gore's effort to involve its partners in "the design and implementation of its program activities" (Medley 2004, 324–25).[10] As EP1 progressed, conservation NGOs and USAID developed a mutually beneficial political relationship that sustained the investments both made in Madagascar.

Building Institutional Capacity

USAID provided significant funding for the development of environmental infrastructure during EP1 (World Bank 2007), and its financial contributions encompassed two major projects aimed at building institutional and legal capacity. The first, entitled "Knowledge and Effective Policies for Environmental Management Project" (KEPEM), was a US$42 million portfolio, including US$33 million of budget support to the Madagascar government, given in exchange for the government's environmental policy reforms, and US$9 million for direct project assistance.[11] Managed by the consulting firm Associates in Rural Development, Inc. (ARD),[12] the project leveraged an impressive number of environmental policy and institutional reform efforts, including the strengthening of the ONE; the passage of a number of environmental laws related to CBNRM, foundation operations, and environmental impact assessment; and technical assistance in the development of an environmental endowment fund, an environmental monitoring system, and environmental impact assessment. Finally, the agency leveraged forest management and revenue system reforms. It helped to establish the forest law of 1997, a forest sector observatory to promote transparency in the permitting process, and a program with the U.S. Forest Service for training on forest inventory, wood value assessment, and competitive bidding as well as technical assistance to forest zoning and the forest commission (ARD 1997; USAID 1994b, 1997d).

The second program, called, "Sustainable Approaches to Viable Environmental Management" (SAVEM), entailed a US$40 million portfolio that continued U.S. investment in integrated conservation and development via two subcomponents. The first supplied technical assistance to the ANGAP in financial management, monitoring, and strategy development through a contract with Tropical Research and Development (TRD) (TRD 1997; USAID 1994b). The second, a cooperative agreement with the NGO Private Agencies Cooperating Together (PACT) entailed two programs. These included (1) small conservation action grants (US$2,000 to US$25,000) to local, government, and grassroots groups for community initiatives in zones around the periphery of national parks (French et al. 1995; USAID 1994b, 1995b); and (2) six large (US$2 million to US$3 million) grants for ICDPs at Amber Mountain, Andasibe, Andohahela, Masoala, Ranomafana, and Zahamena. Funds for a seventh, Isalo, were awarded directly to ANGAP in 1996 (McCoy and Razafindrainibe 1997; PACT Inc. 2000; Swanson 1997; USAID 1994b).[13]

A final program, Tradem, was proposed but never implemented. The negotiations over Tradem reveal the influence exerted by scientists and NGOs on USAID's policies and programs. The goal of Tradem was to generate income and economic benefits from the sustainable trade of natural resource products—an idea that, ironically, reappeared many years later in Ravalomanana's Madagascar Naturally! framework. In preparation for the project, USAID funded the U.S.-based TRD and the Madagascar-based firm Biodev to conduct background studies on the national reforms needed to facilitate trade and its environmental impacts.[14] However, the MFG, which had formed after the St. Catherine's trip (see chapter 3), together with a number of zoos, research institutions, universities, and NGOs, sent a joint letter to George Carner, then-director of USAID-Madagascar. The letter laid out a series of concerns about the proposed program, including its emphasis on sustainable use, the organizations with which it had worked in planning the project, the lack of sufficient scientific population and distribution assessments to ensure sustainable takes, and insufficient NGO involvement. The letter writers complained that "most of us have heard about Tradem primarily through consulting firms seeking our services, which is disturbing."[15] To remedy this oversight, they recommended the provision of "broad consultation with and consensus-building among scientists, conservationists and trade experts both nationally and internationally" before implementing the program. As a result of the controversy USAID abandoned the idea after the planning stage and focused on SAVEM and KEPEM (Freudenberger 2010). The agency's reversal reveals the influence of the previously established scientific / conservation networks (see chapter 3) and the commitment within many conservation circles to isolating nature from humans and avoiding discussions of sustainable use.

Because USAID's funding was contingent on Madagascar's adherence to structural adjustment, when the Madagascar government failed to meet IFI requirements in 1994 the mission underwent substantial funding cuts (Freudenberger 2010). Concurrently, congressional downsizing recommendations in connection with USAID reforms (see chapter 4) closed more than two dozen missions in the mid-1990s (Medley 2004; Smillie 1999). The Madagascar environment program was saved by the biodiversity earmark and the lobbying efforts of CI and WWF in Washington (Freudenberger 2010).[16] The effect of their labor was visible in the report language that accompanied the Senate appropriations bill of FY 1996: "As USAID makes efforts to downsize, it should remain active in regions that are significant for global biodiversity [and] NGOs are often the most cost-effective channels for delivering development assistance" (U. S. Congress 1995).

From CBNRM to Ecoregional Planning

By the mid-1990s various studies had begun documenting problems with ICDPs around the world (e.g., Brandon and Wells 1992; Larson, Freudenberger, and Wyckoff-Baird 1998). In Madagascar critiques of ICDPs specifically pointed to their high expense and the lack of appropriate development expertise within conservation NGOs as well as to the unsuccessful transformation of leadership and insufficient benefits flowing to local communities (Gezon 1997; Hough 1994; Marcus 2001; Peters 1998). They also questioned the impact of ICDPs on biodiversity: a USAID consultant report also concluded, "Madagascar's ICDPs have not provided conclusive evidence that the conservation–development linkage can be made strong enough, with enough people, quickly enough, to have any real long-term impact on the basic problem of continuing biodiversity loss" (Swanson 1997, 4–4).

Amidst this controversy, some conservationists began turning to CBNRM, which advocated that conservation "be pursued by strategies that emphasize the role of local residents in decision-making about natural resources" (Adams and Hulme 2001, 2). Like ICDPs, CBNRM projects aimed to create economic incentives for communities living near and around parks to conserve resources, but they also sought to empower communities to manage these resources themselves rather than to simply benefit from park income and development projects. While USAID was a central sponsor of CBNRM projects, particularly in Southern Africa (Alcorn 2005; Wilson 2005; Wolmer 2003a), it invested in CBNRM in Madagascar relatively late compared to other countries. Prior to the introduction of CBNRM in Madagascar numerous academics and consultants had underscored the importance of empowering local communities there with resource management authority.

After a series of donor-funded CBNRM planning workshops, in 1996 the Madagascar government, backed by the French

Cooperation agency, created a CBNRM program called Gestion Locale Sécurisée (GELOSE) (Secure Local Resource Management). GELOSE transferred natural resource management to rural communities by establishing and guaranteeing local land tenure for subsistence and by gradually granting formal title for land held under customary rights. The first contracts were signed in June 2000 (Bertrand, Horning, and Montagne 2009; Blanc-Pamard, Pinton, and Rakoto Ramiarantsoa 2012; Montagne and Ramamonjisoa 2006). However, GELOSE required a mediator to facilitate the agreement, which slowed its implementation. At the behest of donors such as USAID, which pushed to speed up the process for implementing community management transfers in forests, an additional CBNRM program for forest zones, Gestion Contractualisée des Forêts (GCF) (Contract-based Forest Management), was created in 2001. GCF eliminated the need for an environmental mediator and required a contract only between the community and the DGEF (Bertrand and Ratsimbarison 2004; Healy and Ratsimbarison 1998; McConnell and Kull 2014) (fig. 6.2). Despite the high hopes for CBNRM, donors made insufficient investment in many of these projects to ensure their sustainability as the twenty-first-century rise of more preservationist approaches and ecoregional planning quickly superseded them (Bertrand, Horning, and Montagne 2009; Blanc-Pamard, Pinton, and Rakoto Ramiarantsoa 2012). By 2008 five hundred GELOSE and GCF contracts existed on paper, but many were not well functioning (BATS 2008, 63; see also Hockley and Andriamarovololona 2007; Raik 2007; Resolve 2004).

By the early 2000s participatory conservation had given way to market-based conservation and landscape planning across the globe (Büscher and Whande 2007). In contrast to its late embrace of CBNRM, USAID-Madagascar was at the forefront of the ecoregional planning trend, and it became the centerpiece of the USAID environmental strategy in EP2.[17] The concept of ecoregional planning drew on the science of island biogeography, and critics used it to assert that isolated parks or CBNRM projects were insufficient to conserve biodiversity where habitat fragmentation was a threat. Relying on GIS, they advocated for biodiversity corridors as well as ecoregional and transboundary approaches, which

Figure 6.2. Ampatsy Contract-based Forest Management Community Association. Photograph by Catherine Corson.

broadened conservation planning to a landscape scale (Adams and Hutton 2007; Olson and Dinerstein 1998; Olson et al. 2001; Wolmer 2003b). By using science and GIS as the foundation for centralized and large-scale planning, ecoregional planning corresponded to a reduced focus on local control over natural resources and increased the influence exerted by staff in organizations such as CI, with strong GIS skills (Brosius and Russell 2003; Brosius 2006; Gezon 2000; Wilshusen et al. 2002).

The idea of ecoregional planning appealed to both strict conservationists and advocates of integrated rural development and environmental programs, and the tensions between these views played out in the USAID-Madagascar program. Some USAID officials and contractors saw ecoregions as an opportunity to promote integrated land use planning on a broad scale. A USAID mission official summarized this view as follows: "For me, I think about land-use planning in a broader vision. You have to plan what is going on in all regions, including land-use for forests and agriculture. It integrates rural development and conservation."[18] Likewise, a

USAID contractor underscored the need to work across scales: "Ecoregional practitioners also need to avoid over-promotion of the 'large' or 'big' aspect of the paradigm: the challenge is often working at and linking multiple scales rather than a unique focus on large scale concerns" (Erdmann 2008).

In contrast, many conservationists envisioned ecoregional planning as a means to move beyond ICDPs to connect high biodiversity areas through corridors as well as to protect critical species and ecosystems that were found outside of protected areas. "One of the more critical lessons [from EP1] was that the ICDPs and mini-projects had too narrow a focus in trying to address extremely complex environmental problems" (Cowles et al. 2001, 410). A Global Environment Facility (GEF) workshop held in April 1995 concluded that sizable portions of priority conservation and research areas were located outside of the Madagascar national parks network, and scientists who had attended the workshop argued for immediate action: "This may be the first time there is political will, and the last time there is the biological opportunity. . . . Major new funds are being allocated to Madagascar for corridor approaches (EP2 regional approach) and forests outside of parks" (Hannah et al. 1998, 33–34; see also USAID 1997c). This workshop fed into subsequent biodiversity conservation prioritization exercises, and as the influence of the conservation lobby grew, ecoregional planning became a mechanism for prioritizing conservation at a larger scale and expanding the protected areas network in SAPM.

Nonetheless, USAID staff tried to use the concept to build a more holistic and balanced environmental portfolio. The request for applications issued under the 1997–2002 USAID five-year program for EP2 integrated these views by framing ecoregional planning as a multisectoral approach that would contribute to biodiversity conservation:

> One of the most important re-orientations of EP2 and the USAID environmental program is the decision to expand biodiversity conservation efforts outside the existing protected areas network to add corridors, classified forests, littoral forests, marine / coastal ecosystems, and other

associated landscapes. This decision was made based on the outcomes of a participatory (multi-stakeholder) process to define priorities for biodiversity conservation, where it was concluded that biodiversity should not be addressed solely as a sectoral issue. . . . Since many of the highest priority biodiversity areas fall outside the protected areas system, multiple use strategies for the management of forests and natural resources are needed. (USAID 1997c, 13)

USAID requests for proposals and applications (RFPs and RFAs) for EP2 programs further underscored that EP1 had "paid inadequate attention to agriculture and its clear relationship to environmental degradation" (USAID 1997c, 12). USAID tried to address this shortcoming by funding three main environment programs that tackled ecoregional planning in complementary ways. The first, LDI, was a contract to Chemonics to promote agricultural intensification and commercial benefits from the sustainable use of natural resources and to contribute technical assistance for CBNRM projects (Chemonics International 2004, 3).[19] One of its accomplishments was the Koloharena movement composed of numerous farmer-to-farmer associations and cooperatives.

The second, Projet d'Appui à la Gestion de l'Environnement (PAGE) (Environmental Management Support Project), via a contract to IRG, aimed to mainstream environmental issues into national, regional, and local policy by giving technical assistance to the government on environmental finance, policies, and procedures as well as information dissemination (IRG 2002). It also supported the framework for sustainable protected area financing and the carbon fund pilot project at Makira (Freudenberger 2008).

Finally, the Miray program for Ecoregion-based Conservation and Development entailed a grant to WWF-Madagascar, CI, and PACT to improve the management of critical biodiversity habitats through different ecoregional planning schemes, including (1) PACT-led assistance to AGERAS, the NEAP program to promote decentralized governance, management, and land use planning (see chapter 5); (2) a project entitled Composante Aires Protégées et Ecotourisme (Protect Areas and Ecotourism Component) implemented by WWF-Madagascar, which assisted ANGAP with the management of protected areas; and (3) a project called Ecosystèmes Forestiers à Usages Multiples (Multiuse Forest Ecosystem), managed by CI, which was aimed at improving forest management in nonprotected areas, facilitating land use zoning, and transferring control over forest management to communities (Medley 2004; PACT, Conservation International, and World Wide Fund for Nature 2004; USAID 1995b, 1997b, 1997c, 1998b, 2002c). USAID-funded programs also advanced the development of Plans Régionaux de Développement (Regional Development Plans) and created stakeholder consultative groups for forest management, encompassed in a Comité Multilocal de Planification (Multisite Planning Committee) as well as sponsored a Global Development Alliance partnership with Rio Tinto / QMM in Tolagnaro (Fort Dauphin) that emphasized environmental impact assessment and community consultation (Freudenberger 2008; Seagle 2012). Together, these programs were intended to attack core drivers of biodiversity loss from multiple angles.

In EP3 the USAID mission retained the ecoregional approach and continued to emphasize linkages between rural development and environment programs as it concentrated its efforts in three ecoregions along the eastern rainforest corridor, including the Andasibe / Mantadia–Zahamena Corridor, the Ranomafana–Andringitra Corridor, and the South East Ecological Zone (Tolagnaro) (USAID 2003c).[20] In these strategic zones it again funded another three programs that pursued ecoregional planning in reciprocal ways: Jariala, a contract with IRG to improve the forest management system;[21] Miaro, a grant managed by CI and sub-granted to WCS, WWF-Madagascar, and ANGAP for the improved management of biodiversity habitats;[22] and Ecoregional Initiatives (ERI), a contract with Development Alternatives, Inc., to promote alternatives to slash-and-burn agriculture (USAID 2003a, b, c). As in the case of the EP2 programs, these were intended to complement each other: "First and foremost, there must be a strategic vision for the forest ecosystems and a plan as to how best to achieve that vision. Second, within these forest ecosystems, there must be core protection zones for critical biodiversity habitats, which fulfill the need to protect priority natural resources and ecological processes. Around these core areas will be sustainable use zones, which may be privately, publicly, or locally managed" (USAID 2003c, 11).

Impediments to Integrating Agriculture and Environment

To encourage USAID contractors and grantees in the three programs to collaborate, the USAID RFPs and RFAs in 2003 included the following note to applicants: "A high priority will be placed on effective communication, coordination, and collaboration between these different components and with other USAID Strategic Objectives. Activities will be integrally linked to other partners and activities in order to achieve results" (USAID 2003c, 37). As a mechanism to bring this about, USAID introduced ecoregional alliances in 2003. Comprised of all USAID implementing organizations and partners, they included nonenvironmental programs such as Title II food aid, democracy, health, and business programs (BATS 2008).[23] Based in the regional capitals and the national capital, the alliances held monthly meetings of all environmental partners or partners with environmentally related programs (or both).[24]

After SAPM was introduced, however, the managers of the projects often worked at odds with each other in the park expansion effort. Through the Miaro program USAID was a core bilateral donor to the SAPM process. However, its Jariala project assisted the forest zoning initiative, which aimed to zone forest areas not just for conservation but also for commercial forestry and fuelwood supplies.[25] Jariala also assisted the mining and forest commission, which coordinated activities between MinEnvEF and the Ministère de l'Energie et des Mines (Ministry of Energy and Mines), including overlapping conservation and mining sites (see chapter 7) (Repoblikan'i Madagasikara 2004a, b; USAID 2003c). Finally, the USAID ERI project aspired to increase small-scale agricultural productivity, to develop regional planning for the forest corridors, and to promote CBNRM (Erdmann 2010; USAID 2003a).[26] However, faced with pressure to implement the high-profile SAPM initiative rapidly (see chapter 7), conservation goals often took priority.

There were mixed reports on the effectiveness of the alliances: while regional USAID contractors and grantees in each of the three USAID ecoregions cited their benefits as coordination of activities and data sharing, others reported that the alliances did not effectively coordinate activities in the field.[27] One development NGO representative responded by saying, "[USAID] does a great job in trying to force that linkage by trying to create the ecoregional initiative and by putting the requirement in the RFPs, but at the end of the day, it is environmental organizations doing environmental programs and others (development organizations) are working in different areas."[28] Staff members of USAID, regional contractors, NGOs, and government agencies argued that there was "too much emphasis on environmental issues, like SAPM, that Title II and Bamex (other USAID programs) are not involved in."[29] A conservation advocate confirmed this assertion from the other side: "The [ecoregional] alliance is a mechanism for making clear the goal is biodiversity, not development. All other money that is going to other sectors should be moving toward conservation."[30] As another attempt to build inter-sectoral coordination, in 2005, the Madagascar mission adopted the NWP framework (see chapter 4) and added a health and nutrition component to it, making it Nature, Health, Wealth, and Power (NHWP). Its health and environment programs ranged from family planning to basic health, and its NHWP program became a model for USAID (Freudenberger 2010).

In 2008, in a "stock-taking exercise" aimed at reviewing USAID's experience in environment and rural development, numerous USAID staff and partners emphasized not only the need to take a more comprehensive approach to environmental issues in Madagascar but also the importance of addressing broader drivers of degradation. Some of the programmatic shortcomings identified were the need to offer long-term institutional support (Erdmann and Rakotodrabe 2008), to fund rural development and build institutional capacity (Andrianandrasana et al. 2008), and to invest in local community capacity building (Razafitsotra et al. 2008), including benefits that go beyond the monetary (Rajaspera et al. 2008). A subsequent view of the ERI program recommended that "given the continued poverty of Madagascar's rural population, sustainable development, not biodiversity conservation, should be the driver of broad-scale development and conservation initiatives. . . . Moreover, agriculture as the foundation of the population's livelihood strategy must continue to receive unrelenting attention" (Erdmann 2010, 391).

However, despite the mission's initial emphasis on connecting agricultural productivity and biodiversity conservation—and even as mission staff and partners stressed the importance of addressing the drivers of environmental degradation and launched initiatives such as NHWP and the ecoregional alliance—their efforts to pursue more holistic programs faced a number of impediments. First, staff had to work within the framework of the mission's overarching goal of promoting economic growth and the guidelines for using the biodiversity earmark, which funded almost 100 percent of the USAID-Madagascar environmental program over its lifespan (USAID 2008). This discouraged integrated agriculture and environment programs as well as initiatives that challenged economic growth based on extraction of natural resources.[31]

As Freudenberger (2010) reports, "USAID institutional and bureaucratic systems and, especially, sectoral divides, made it extraordinarily difficult to implement a truly holistic approach on the ground. Persistent attempts to overcome these difficulties at the Mission and field level were only partly able to compensate for the lack of flexibility and the near impossibility of creating projects that were as multi-sectoral as the problems they were trying to address" (29). In particular, given the difficulty of demonstrating immediate, direct, and measurable impacts on biodiversity, projects that addressed more indirect governance, economic, and social aspects of biodiversity loss could not access the earmarked funds.[32] The earmark's discouragement of cross-sectoral, holistic, or creative programs was furthered by the increased emphasis by headquarters during this time period on reporting results and the decline of economic growth and democracy / governance funds, which had less congressional support than the health and environment sectors.

Agriculture and environment programs also competed for funds. A former USAID official remembered how this tension played out in Madagascar: "In Madagascar, there was a decision . . . [that] the biodiversity money was going to be spent on agriculture and sustainable agriculture, under the argument that you need to focus outside the protected areas on making the agricultural landscapes more sustainable. This was both so people stay there and don't go into the parks, and also because you have important natural resources in those agricultural landscapes. But it was always perceived as like a zero sum. What the agriculture folks get, the biodiversity folks don't get."[33]

As USAID's biodiversity emphasis grew, the Madagascar mission shifted its articulated strategic goal for the environmental program from natural resource management to biodiversity conservation. The goal for the 1993–98 environmental program was to "Increase Sustainable Natural Resource Use in Target Areas" (USAID 1992a). However, from 1998 onward the program's overarching goal became to "Conserve Biologically Diverse Forest Ecosystems" (USAID 1997d, 2002c). A mission official observed that, when planning for the latter program, they feared that, in the context of overall USAID budget decreases, they might not get environmental funds if they did not make biodiversity explicit so they could use biodiversity earmark funds. During the development of the EP3 strategy, Washington-based staff similarly advised that the mission keep biodiversity conservation as its primary objective so that it could use the earmarked funds.[34] This became especially critical during the George W. Bush administration, when the biodiversity earmark was one of the few secure sources of environment program funds. Thus even as the mission continued to fund sustainable natural resource use programs, it defended them in terms of their contributions to forest biodiversity conservation.

Furthermore, projects funded by the biodiversity earmark were supposed to meet the guidelines that had been drawn up by staff in USAID, NGOs, and Congress. These guidelines emphasized that any project funded with earmarked funds must (1) have an explicit biodiversity objective; (2) be identified based on an analysis of threats to biodiversity; (3) monitor biodiversity indicators; and (4) have the intent of positively impacting biodiversity in biologically significant areas (USAID 2007b). Such strict guidelines discouraged investments in rural development, sustainable agriculture, and economic growth, privileging instead projects that were more easily defended as focusing on biodiversity, such as expanding parks.

In the same vein, because the overall strategy for the USAID mission as a whole has been to promote economic growth, the environment program had to be justified in terms of its relationship to economic growth.[35] For example, the strategy for 1998–2002 argued that the environmental strategic objective would contribute to the achievement of USAID-Madagascar's overall development assistance goal of "Broad-based Sustainable Economic Growth" by "conserving the resource base upon which the productive capacity of the next generation of Malagasy and the sustainability of the country's future economic growth depend" (USAID 1997d, 68). Then, the 2006–7 Washington-based reforms subsumed USAID under the State Department and situated environment programs under the broad category of economic growth programs, which were supposed to "provide financial resources and technical assistance to promote broad-based growth" (U.S. Department of State 2007a). This created additional political pressure to align environmental programs with the political and economic goals of the United States and to solicit political backing from diverse actors in Washington.

Saved by the Conservation Lobby

Reforms emanating from Washington-based agencies forced alignment among U.S. foreign aid, foreign policy, and defense policy, introduced more top-down control over mission activities by requiring reporting against centralized indicators, and gave U.S. ambassadors a more active role in deciding foreign aid priorities.[36] These changes reinforced the need both to justify programmatic relationships to overarching strategic goals and to build informal networks in Washington. In accordance with the Government Results and Performance Act of 1993 and accompanying agency reforms, the mission had to measure the impact of its activities against quantitative indicators (Medley 2004). Reforms made in 2006–7 required missions to reduce their operating expenses, including utilities, buildings, and salaries,[37] as well as to report their accomplishments against indicators determined by headquarters. A mission officer confided, "We just got standardized indicators from F [the State Department's Office of U.S. Foreign Assistance Resources] that we have to report on that have nothing to do with what we have been doing and that we have not been tracking."[38] The audit of the mission's biodiversity program conducted in 2008 by the USAID Office of the Inspector General critiqued the mission's incomplete assessment of its accomplishments against target indicators, even as many of the factors influencing its ability to meet the indicators were out of the agency's control. The audit recommended that the mission improve performance reporting against indicators, verify data from grantees and contractors, and ensure that its grantees and contractors comply with USAID branding requirements, which require missions to assure that the USAID logo is on all vehicles, project signs, and other funded materials (USAID 2008). This emphasis focused attention not on doing effective development but on reporting and advertising it. The mission also had to take on the increased reporting with fewer staff.

While I was doing field research the Madagascar mission "had to reduce staff by one-third,"[39] as part of a "harmonization exercise" to better align the number of staff with mission budgets, and to adjust programs in response to cuts in economic growth and democracy funds.[40] By 2008 budget cuts and reductions in force had reduced full-time mission staff to three, which a 2008 evaluation conceded was "not sufficient to effectively manage all program activities and maintain the valuable leadership role in Madagascar's environmental community" (BATS 2008, 103). The continual reductions in direct-hire staff necessarily dispersed program and policy management to contractors, NGOs, and temporary employees, which undermined USAID's capacity to conduct long-term strategic planning and investment.

The combined reduction of the state and the rise of the nonprofit sector under neoliberal reforms in both the Clinton–Gore and George W. Bush administrations reinforced the need to maintain strategic, often informal, alliances with Washington-based NGOs who could use their leverage in Congress to ensure continued funding streams for the environment. Thus instead of decentralizing control over mission activities, as neoliberal ideology might presuppose, the reforms reinforced upward accountability. Mission staff stressed that their client on a day-to-day basis became Washington rather than those people affected by the programs.[41] As the conservation lobby based in Washington turned into a critical ally, a focus on biodiversity conservation became a strategic maneuver to ensure continuous funding. Referring to the mid-1990s reforms, a USAID mission official summarized, "Madagascar is a biodiversity hotspot—that's what we sell to Washington. . . . If it was not for [biodiversity] hotspots,

Madagascar would be closed with the other twelve [USAID] missions that were closed in Africa." In fact, the USAID environmental program persisted in Madagascar in the wake of these reforms and the Bush administration's general disinterest in environmental issues precisely because USAID mission staff had successfully built strategic political relationships with Washington. The official continued, summarizing their strategy: "We try to have the DC biodiversity team come here. . . . Every two years we have a partners meeting in DC, [and a mission official] goes to make a presentation. . . . We also try to have congressional members come. . . . We send them with our ambassador to see Andasibe or Fort Dauphin. So we work with NGOs, our partners, Congress. . . . As a result, we haven't really had budget cuts, like other strategic objectives have."[42] In fact, I took my first trip to Madagascar as a headquarters USAID officer invited to visit by the mission's environment officer.

In Washington, conservation NGOs lobbied not just for biodiversity conservation but also specifically for the USAID Madagascar program. The ICC and its individual partners were instrumental in raising congressional and private awareness about Madagascar's biodiversity program. In particular, CI built a reputation for both effective biodiversity lobbying and relations with the Madagascar government as well as its focus on high-level fund-raising, often at the expense of effective field-based intervention. CI officials had close relations with the former World Bank president James Wolfensohn and lobbied for World Bank investment in Madagascar's conservation program,[43] for example, in regard to the Madagascar Biodiversity Fund.[44] At the same time, CI staff took various board members to Madagascar, including Jeffrey Katzenberg, the chief executive officer of the motion picture company DreamWorks Studios, Gordon Moore of Intel, and Rob Walton from Walmart as well as two chairmen of the ICC, Congressmen Tom Udall and John Tanner.[45] Like the trips discussed in previous chapters, these helped to build widespread congressional support for conservation in Madagascar.

The ICP and ICCF also organized a number of events that nurtured personal connections between members of Congress, the conservation NGOs, and Madagascar government officials. These included the screening of a film funded by the USAID-Madagascar mission, "Madagascar: A New Vision," for members of the U.S. Congress.[46] Sponsored by the conservation NGOs, the Madagascar minister of environment, water, and forests came to Washington for the screening in order "to demonstrate Madagascar's commitment to biodiversity."[47] According to the ICCF website, "This event highlighted the country of Madagascar for their historic commitment to triple the size of their national park system. . . . Members who visited Madagascar shared some of their experiences from touring the island's unique flora and fauna" (ICCF 2008). Later in the year the ICC organized a breakfast for conservation NGO representatives based in Madagascar with five members of Congress.[48] Finally, in 2009 the ICCF hosted a congressional briefing with CI, WCS, and WWF-U.S. on the negative effects of discontinuing environmental funding following the coup in Madagascar, and in 2010 Lisa Gaylord, then the WCS director of global program development but previously the long-term USAID Madagascar environmental officer, briefed congressional staff in an event hosted by ICCF titled "How Biodiversity Conservation Accomplished U.S. Priorities on Improving Health and Stopping Hunger [in Madagascar]." These events circumvented the official channels USAID staff, as federal employees, had to use and solidified informal relations among NGOs and congressional members.

As the NGOs pushed to preserve the USAID biodiversity program in the face of proposed cuts, they both drew on and *produced* the Madagascar mission's exemplary status. When, in 2005, the MCC awarded its first grant to Madagascar many people feared the award would lead to a reduction in funds for the USAID biodiversity program, as USAID had announced its intention to reduce funds in MCC-recipient countries.[49] At the behest of the conservation NGOs, the House ICC chairmen—Congressmen Clay Shaw, Ed Royce, Udall, and Tanner—wrote to Secretary of State Condoleezza Rice and USAID Administrator Natsios, "Some may see the US$110 million from the MCC as an opportunity to reallocate USAID resources elsewhere," they said, "[but] USAID's conservation investment in Madagascar is a powerful example for how biodiversity conservation can support economic growth, community development. . . . We urge you to ensure that it receives sustained support" (Shaw et al. 2005). In March 2007 WWF-U.S.'s vice president for government relations, Jason Patlis, submitted testimony on behalf of WWF-U.S., CI, TNC, and WCS to the House Appropriations Committee regarding USAID's budget for FY 2008, which argued that Congress should preserve the mission's biodiversity program in the face of budget cuts under the foreign aid reform of the Bush administration: "We have serious concerns with the implementation of the new [foreign aid reform] strategy. Madagascar is one of the world's richest countries in terms of biodiversity, and its government has identified the environment as a critical sector for the country's social and economic development—and yet USAID has slashed its environment funding for that country by 40 percent" (Patlis 2007).

All of these events nurtured the development of critical, informal, political relationships among USAID, Madagascar government officials, the conservation NGOs, and key members of Congress. While USAID officials had to contact congressional members and staff through official channels, NGO representatives and the Madagascar government officially could speak more informally to members. They depicted Madagascar as an exceptional biodiversity haven. The result, according to several interviewees, was that mission officials felt pressure to give grants to CI, WWF, and WCS so as to maintain the necessary support in Washington for its Madagascar program.[50] One person stated bluntly, "USAID couldn't not give money to CI/WWF. They would go straight to the Hill [Congress]."[51] Another summarized, "USAID wouldn't be in Madagascar without WWF/CI lobbying in DC. Then USAID in turn funds them to do what they want in Madagascar."[52] The resulting political alliance safeguarded both the focus on strictly defined biodiversity conservation as pursued through expanding protected areas and the long-term stability of the USAID mission.

A Mutually Beneficial Relationship

By moving between policy-making sites in the United States and Madagascar, one can see how USAID's environmental policy and programs were developed, implemented, and continually negotiated via transnational relationships among state and nonstate actors. The historical analysis of USAID's environmental program in Madagascar reveals not only the rising power of the conservation NGOs but also how they and the mission developed a mutually beneficial relationship that evolved through progressive conservation trends. American scientists and conservationists began as external advocates, organizing to draw global attention to Madagascar's biodiversity, and USAID relied on them, first, for information and then to implement ICDPs. While ICDPs helped to bring the large conservation NGOs into conservation policies, CBNRM marginalized them in favor of local managers. The emergence of ecoregional planning in the twenty-first century then revived them as major players not only in managing conservation but also in deciding where it should be focused. As USAID underwent neoliberal reforms in the 1990s that downsized the agency and encouraged the involvement of nonstate actors in its policy and programs, it began directing increasing funds through contractors and NGOs. Then as U.S. funding scaled up, it facilitated the transformation by conservation NGOs from their positions as researchers and external advocates to financial conduits and ultimately part of the political and bureaucratic infrastructure constructing the Madagascar environmental program.

In the context of federal government reforms and declining environmental funds overall, the biodiversity earmark was essential in ensuring environmental funding, but it also dissuaded investment in integrated, holistic conservation, as the mission had to comply with guidelines from Washington for implementing the biodiversity earmark. This discouragement was reinforced by bureaucratic factors such as the fact that agriculture and environment programs compete for the same funds. The mission's overt prioritization of biodiversity and its emphasis on national parks was a strategic maneuver to assure a continued stream of environment funds guaranteed by congressional interest in biodiversity. The mission nurtured Washington's support for its programs by bringing headquarters staff to visit, by briefing the ICC through NGO-organized events, and by building alliances with conservation NGOs, who also solicited backing from members of the Congress, moviemakers, and foundation leaders. In turn these congressional caucus leaders pressured Congress to protect environmental funding to Madagascar in the face of overall foreign aid cuts. As a result, U.S. environmental foreign aid flourished in Madagascar even as USAID cut back its environmental programs elsewhere.

In July 2009, after the political crisis, USAID's program in Madagascar was suspended and US$7 million in funds reprogrammed. Freudenberger (2010) notes that Congress has given biodiversity conservation programs the status of "notwithstanding authority," which allows them to be managed independently because of their global importance. The Obama administration decided not to invoke this authority after the political crisis, despite heavy lobbying by conservation NGOs and others. This decision reveals the importance of attending to the constantly shifting nature of power relations among USAID, the Department of State, Congress, and the conservation NGOs, and it underscores that environmental issues remain marginal in American politics. That said, U.S.-based conservation NGOs continued to operate in Madagascar, and even as USAID pulled out after the coup they successfully lobbied the Madagascar interim government to increase the protected area expansion program (see chapter 7).

CHAPTER SEVEN

Accountability and Authority in Conservation Politics

We can no longer afford to sit back and watch our forests go up in flames. This is not just Madagascar's biodiversity, it is the world's biodiversity. We have the firm political will to stop this degradation.

—FORMER PRESIDENT OF MADAGASCAR MARC RAVALOMANANA at the Fifth World Parks Congress in 2003

THE PLEDGE MADE BY former president Ravalomanana in 2003 at the Fifth World Parks Congress in Durban, South Africa, to triple Madagascar's protected areas represented the culmination of decades of work by foreign and Madagascar government officials, scientists, transnational NGOs, and others to save biodiversity in one of the world's top five hot spots.[1] The initiative, first known as the Durban Vision and later termed SAPM, aimed to triple the country's protected areas from 1.7 million to 6 million hectares in five years in order to meet the IUCN target of protecting 10 percent of every biome. It became a major focus of the final phase of the fifteen-year

NEAP. By December 2010 the 6-million-hectare goal had been surpassed, with protected areas and sustainable forest management sites covering 8.7 million hectares and an additional 10.5 million more identified as potential sites (Repoblikan'i Madagasikara 2010a, b). Then, at the Sixth WPC in 2014, Ravalomanana's successor, President Hery Rajaonarimampianina, agreed to continue the expansion by tripling Madagascar's marine protected areas by 2020.

Advocates publicized the initiative as a groundbreaking way of establishing and managing parks. A USAID report called it "a major shift in how ... protected areas are understood in Madagascar" (USAID 2007a, 22). Senior Vice President John Robinson of WCS asserted, "This commitment recognizes the importance of parks as a way to both protect biodiversity and to promote sustainability and national development in the rural landscape. Madagascar is clearly leading the way towards this vision by promoting long-term partnerships with all sectors of civil society" (CI 2003).

In his announcement Ravalomanana underscored that the new protected areas would adhere to IUCN guidelines, which delineate seven categories of parks ranging from those that prohibit human entry to those that allow sustainable use; encourage consultation with potentially affected local populations; and promote comanagement with a variety of public and private entities, including local communities (Dudley and Phillips 2006; IUCN 2004).[2] In line with these recommendations and in recognition that neither the forest nor parks service had the capacity to manage additional parks, policy makers underscored that a multiplicity of actors—such as NGOs, the private sector, and local communities—could manage the new parks (Commission SAPM 2006). They also emphasized that the new protected areas would involve communities to a greater extent than previous ones. Numerous policies stressed the need to consult with potentially affected communities (e.g., Borrini-Feyerabend and Dudley 2005a; Commission SAPM 2006; Repoblikan'i Madagasikara 2005c; World Bank 2005), and donors and conservation NGOs touted the program's novel approach.

My research revealed limited community involvement in the initial establishment of the protected areas as well as the appropriation of resource rights previously granted to local communities. While the high-profile announcement successfully mobilized funding for biodiversity conservation, the foreign and domestic political attention paid to SAPM undermined efforts to consult with rural communities, instead reinforcing upward accountability to national politicians, foreign aid donors, and transnational conservation NGOs based in Antananarivo. Furthermore, although the president of Madagascar made the official decision to expand the nation's protected areas, nonstate, often foreign, actors shaped the boundaries, resource rights, and decision-making authorities associated with the new protected areas, thereby consolidating their ability to determine land and resource rights through SAPM. The state became a vehicle through which numerous nonstate entities could expand their control of and authority over Madagascar's forests. Finally, SAPM reproduced many of the historical conflicts over Madagascar's forests that led to its current biodiversity crisis by reclaiming state authority over forests, separating conservation and production zones, protecting minimal usage rights for villagers while restricting their commercial activities, and accommodating commercial forestry and mining interests.

This chapter forms the final "point of entry" into my analysis of USAID's role in the Madagascar environmental program.[3] I examine how negotiations among the Madagascar state, multilateral and bilateral donors, transnational conservation organizations, mining companies, and community leaders shaped SAPM's design and implementation. In tracing the steps through which the initial protected areas were established, I begin by discussing the history behind, rationale for, and negotiations around the announcement in 2003. I explore the contestations and compromises that took place in the processes of mapping, classifying, and designating resource uses in new areas. Focusing on the Ankeniheny–Zahamena and Fandriana–Vondrozo biological corridors in Madagascar's eastern rainforest, I analyze why only limited consultations with rural populations took place when these areas were established as temporary parks in 2005 and 2006, respectively. Finally, I discuss the potential for the initiative to not only produce paper parks but also to catalyze increased deforestation.

As in other chapters, I focus specifically on U.S.-based organizations. USAID and Washington-based conservation NGOs as well as the World Bank were political and financial forces behind both the Durban parks announcement and its implementation.

Many attributed the rapid implementation of the program and the restrictions on resource uses to American influence. USAID officials and U.S.-based NGOs from Madagascar attended the Fifth World Parks Congress as part of the delegation from Madagascar. USAID funded the development of Madagascar's parks legislation, and it was the primary bilateral donor for SAPM in EP3, particularly through the Miaro program. USAID contractors, through the Jariala and ERI projects, were also involved in SAPM negotiations, pushing for the inclusion of sustainable forest sites and community consultation, often in opposition to Miaro. Finally, USAID provided intellectual leadership to the SAPM process as cochair of the NEAP Comité Conjoint and the SAPM technical committee, and CI, WWF, and WCS brought their own additional funding and intellectual contributions as active participants in SAPM working groups. Thus representatives of these primarily U.S.-based organizations were instrumental in shaping SAPM's design and implementation, albeit not always in a unified direction (see chapter 6).

The Decision to Expand Protected Areas

The Malagasy delegation to the Parks Congress in 2003 included government officials from the Madagascar MinEnvEF and ANGAP. Yet, as is common in delegations to international environmental meetings, it also contained NGO and multinational organization delegates, including foreign organizations such as the British Durrell Wildlife Conservation Trust, the Madagascar-based conservation organizations Fanamby and Institut pour la Conservation des Écosystèmes Tropicaux (Institute for the Conservation of Tropical Ecosystems), and the transnational conservation NGOs CI, WWF-Madagascar, and WCS as well as representatives from USAID and the World Bank.

A subgroup of the official delegation formulated President Ravalomanana's pledge to triple his country's protected areas. Prior to the announcement, a small group of NGO representatives and government officials gathered in a hotel room to debate the details of an announcement they hoped would attract international attention and foreign aid as well as expropropriate forest service lands for conservation.[4] As a member of the subgroup recounted, "We were focused on how to make an announcement, which would strike the world. . . . There were three or four people: we couldn't decide what to put: two million or three million. . . . But the deciding factor was that tripling is a little less than 10 percent of the surface of Madagascar."[5] Another member confirmed that "we had ratified the Convention on Biodiversity, which advocates for 10 percent. We had only 3 percent. That's [how] we persuaded the president to make this declaration in Durban."[6] As one delegation member recalled, "The president saw [the SAPM announcement] as a marketing opportunity for Madagascar: a way of making Madagascar number one for something."[7] The final announcement proposed tripling Madagascar's protected areas to meet the 10 percent target, and according to a senior Madagascar government official it did attract additional foreign aid from both current and new donors.[8]

Interviewees often cited three studies as being critical in convincing the former president and his advisors to increase protected areas. The first, a USAID consultant report, showed that forest loss was less in both protected areas and areas more generally where USAID had intervened (Hawkins and Horning 2001). The second, a forest cover map produced by CI, revealed an average 8.6 percent decline in forest cover across Madagascar between 1990 and 2000 and indicated that rates of deforestation were lower in protected areas (CABS 2007; Gorenflo et al. 2011).[9] Finally, a third study, conducted by the World Bank economist Jean-Christophe Carret, used cost-benefit analysis and willingness-to-pay calculations to demonstrate that the protected area network could generate significant economic benefits for Madagascar in terms of tourism, biodiversity values, and watershed protection (Carret 2003). Just as the efforts to quantify the economic costs of environmental degradation convinced Madagascar government officials to embrace the NEAP in the late 1980s, this economic analysis helped to convince skeptical government officials that expanding protected areas would be economically beneficial.[10] As a senior conservation NGO leader said, "In the cabinet the president is not necessarily listening to the minister of the environment. . . . He is listening to the minister of finance [and the] prime minister. And when an essential report comes to his desk demonstrating the economic importance of the forest that is left in Madagascar—that it has a better use being protected than being logged, or being slash and burned—this is where you make progress."[11]

Having made such a public announcement, however, Ravalomanana faced substantial pressure to implement it successfully. Through an association first entitled the Durban Vision Group and later called the SAPM Commission, an alliance of Madagascar government officials, foreign aid donors, consultants, and national and transnational NGOs based primarily in Antananarivo scrambled to make the parks announcement a reality. The commission's official executive steering committee included only Madagascar government officials (Pollini 2007), but the environmental officer for USAID-Madagascar and the director general of ONE led its technical secretariat, which administered the specifics of the program and coordinated several working groups that oversaw aspects of SAPM's implementation (Commission SAPM 2006).[12] Through these working groups Madagascar government representatives, foreign aid donors, consultants, scientists, and national and transnational conservation NGOs undertook functions that would otherwise have been state responsibilities. They drafted, shared, marked up, and finalized guidance, policies, laws, and maps that were ultimately issued officially by the Madagascar government. One of these was the prioritization process for new conservation areas.

The idea to expand the number of parks was not new. Years of advocacy by scientists and policy makers to expand Madagascar's park network to include representation of all the country's major ecosystems preceded the Durban announcement. In addition to Mittermeier's Conservation Action Plan (Mittermeier 1986), Nicoll and Langrand (1989) had proposed the expansion of Madagascar's protected area network to conserve Madagascar's biodiversity more comprehensively. In April 1995 the GEF scientific workshop in Madagascar concluded that a sizable portion of high-priority areas for conservation and research were located outside of the existing parks (Hannah et al. 1998); this conclusion prompted subsequent efforts to expand the park network. USAID funded a park expansion planning exercise that culminated in a National Protected Area Management Plan entitled PlanGrap that proposed additional sites for protected status (ANGAP 2003). An ensuing exercise in biodiversity priority setting entitled Réseau de la Biodiversité de

Madagascar (REBIOMA) (Madagascar Biodiversity Network) in turn developed maps to help policy makers identify conservation priorities by overlaying GIS layers of estimated animal and plant species ranges with estimated threat levels (Kremen et al. 2008; Randrianandianina et al. 2003). The SAPM working group on prioritization used criteria such as the number of species and habitats that would be conserved in a given area—emphasizing rare and endangered species—to arrive at the highest priorities (Rasoavahiny et al. 2008).

These initiatives, reflecting the interests and expertise of the donors, NGOs, and scientists engaged in these endeavors, focused almost exclusively on biodiversity, and, limited by data availability, they concentrated on terrestrial vertebrates. None of them incorporated socioeconomic data, such as where people lived or what environmental resources they used for their livelihoods. Thus when the SAPM process began, policy makers had considerable information about forest fauna but very little information about how rural peasants used forest resources. While there were regional and local efforts to map community conservation sites and differentiated land uses, the resulting national protected area maps were based exclusively on biodiversity priorities, and, in effect, they erased the inhabitants, their livelihoods, and their existing community-managed areas from the targeted landscapes (see also Tucker in Kull et al. 2010).

While increasing the number of parks was one of the overall objectives of EP3, the Durban announcement sped up the process of park expansion, increased its extent, and brought international and national political pressure to bear on it. As one senior Madagascar governmental official recounted, "The original plan was to increase protected areas by 6–7 percent with no limit on the time frame, but then [the conservation] organizations pushed for 10 percent in five years."[13] For many this was the opportunity they had been waiting for: a foreign conservationist who had worked in Madagascar for decades emphasized, "We worked very hard in difficult conditions making progress, bit by bit, and suddenly [it is a] dream come true."[14] Yet the pressure it brought catalyzed intense conflicts among scientists, commercial interests, state officials, NGOs, and foreign aid donors to decide how to manage Madagascar's resources (Corson 2011).

Struggles to Control Madagascar's Resources

Immediately after Ravalomanana's announcement mining and timber extractors, in an effort to protect their interests, rushed to exploit forestlands that might become national parks. They used legal and illegal means, both by working through the formal permitting process and by exploring without permits.[15] The discovery by French donors during an overflight of a seventeen-kilometer-long road constructed for a joint Malaysian/Malagasy commercial forest extraction project in Ambohilero Classified Forest in the eastern rainforest corridor galvanized SAPM advocates into retaliatory action (Pollini 2011).[16] The government halted new mining and timber permits in any *potential* new protected areas until the parks could be established. Scientists who had been working on biodiversity prioritization were suddenly catapulted into the limelight, as policy makers scrambled to put together a map that would show the locations of new protected areas in which mining and timber permits should be banned. The resulting *arrêté* (order), issued by the government in October 2004, suspended all mining and timber permits for two years in any area proposed for protection in a new conservation site, as defined by the accompanying map (fig. 7.1) (Repoblikan'i Madagasikara 2004c). In their haste to put together the order, however, policy makers relied primarily on a GIS layer developed to prioritize lemur habits.[17] This rushed translation of science into policy brought new meaning to the frequent juxtaposition of lemurs versus people in critiques of Madagascar's environmental agenda (e.g., Peters 1998; Simsik 2002). This new territory included the vast majority of Madagascar's forests, and the order was to be renewed every two years. Figure 7.1 shows newly proposed protected areas as well as previously established protected areas.

The hastily constructed map frustrated both mining and conservation advocates with its inaccuracies, and government agencies, NGOs, and mining companies began negotiating. Representatives of mining firms complained that the map creators had used inaccurate GIS data that included unforested, low-biodiversity areas. Conservationists retorted that certain high-priority biodiversity areas had been excluded, which left these areas to the mercy of miners. Thus began a series of informal and official negotiations between

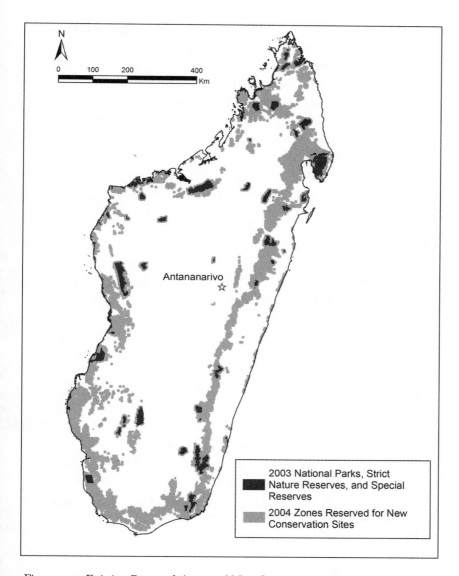

Figure 7.1. Existing Protected Areas and New Conservation Sites in Madagascar as of 2004. Credit: Catherine Corson. Data sources: The country outline is from ESRI, DeLorme Country data; 2003 parks data is from IUCN and UNEP-WCMC (2003). The World Database on Protected Areas (WDPA) [On-line]. Cambridge, UK: UNEP-WCMC. Available at: www.protectedplanet.net [Accessed (July 23, 2003)]; and the 2004 protected areas data is from Rebioma. These data were used in the map accompanying Arrêté Interministériel n° 19560/2004/MinEnvEF/MEM. Available at http://www.rebioma.net/ [Last access June 6, 2014].

transnational conservation NGOs and mining companies. In some cases NGO representatives and mining agents made joint field visits to contested areas to ground reference the map and negotiate new boundaries.[18] In other areas large mining companies like Dynatec and QMM agreed to set aside alternative areas as private reserves or to contribute to the donor-funded Biodiversity Trust Fund so as to have a net positive effect on biodiversity (Sarrasin 2006a; Seagle 2012).

These informal dialogues between conservation and mining interests paralleled the official negotiations that took place under the Comité Interministériel des Mines et des Forêts (Interministerial Mining and Forest Commission), which was established in 2004 to coordinate between the Ministry of Energy and MinEnvEF in order to mediate conflicts where mining permits had been granted in environmentally sensitive areas (Repoblikan'i Madagasikara 2004a, b). Prior to the 2004 order, the Bureau du Cadastre Minier (Office of Mining Registration) had issued mining permits in many of the areas claimed for SAPM, and it continued to grant mining permits in these areas after 2004.[19] In their analysis of the national overlap between mining concessions and protected areas, Cardiff and Andriamanalina (2007) found that mining concessions, primarily for research permits, overlapped with 33 percent of the surface area that had been slated for protection in 2005 and 21 percent of that planned for 2006. In the Fandriana–Vondrozo corridor, regional SAPM committee members discovered that by 2006 mining permits had already been granted in four-fifths of the areas slated for protected status.[20] The 2006 order that established the Fandriana–Vondrozo corridor as a new protected area attempted to resolve this conflict by stating that permits granted before the October 2004 prohibition of mining and logging in new protected areas would be honored, although owners would have to conduct environmental impact assessments before proceeding (Repoblikan'i Madagasikara 2006b).

In 2006, as the government prepared to renew the 2004 two-year protection order for another two years with a revised and improved map that incorporated the outcomes of these negotiations, the DGEF began protesting that SAPM's overwhelming focus on conservation ignored the need to protect timber and fuelwood supplies. An international conservation representative retorted, "SAPM has taken all the forestry surface area, so there is nothing left for commercial use,"[21] and a former DGEF official summarized, "The forestry administration is terrified that if this expansion of the protected area turns out better, it will be the ANGAP who will have all the forests of Madagascar."[22] With assistance from USAID Jariala—a forest sector reform project managed by the American contractor IRG—the DGEF proposed that the SAPM maps also include "sites of sustainable forest management," entitled KoloAla,[23] that would promote development through sustainable use of timber and nontimber forest resources (Burren 2007). The result was the addition of sites of sustainable forest management in the renewal of the 2004 order (fig. 7.2). SAPM thus revitalized the historical tension between conservationists and the forest service that had been initiated by the creation of ANGAP in 1990, when donors, including USAID, poured money into ANGAP rather than into the forest service, producing dramatic differences in salary and other physical amenities (see chapter 3) (Gezon 2000; Hufty and Muttenzer 2002; Montagne and Bertrand 2006).

Through these various means, branches of the state, NGOs, and mining companies negotiated the boundaries and rights associated with the new parks. Their efforts minimized the extent to which the expansion of protected areas interfered with the rapidly expanding mining industry while simultaneously limiting use rights by local residents (Corson 2011).

CBNRM projects remained "off the official maps," and as a result there was no spatial recognition of the fact that many of these lands had already been transferred to local communities to manage. Neither the 2004 nor the 2006 map included preexisting CBNRM sites such as those created through GELOSE and GCF. Some negotiators justified this omission by citing the fact that the boundaries of most of these projects had not yet been geo-referenced. In 2006 the majority of CBNRM projects developed in Madagascar still existed primarily on paper and had received little follow-up since they were first signed (Hockley and Andriamarovololona 2007). While there were regional attempts to incorporate community management transfers, the national maps that were developed to demarcate new protected areas did not include many preexisting CBNRM sites for several intertwined reasons.[24] In many cases paper maps showing CBNRM sites existed, but the spatial data used

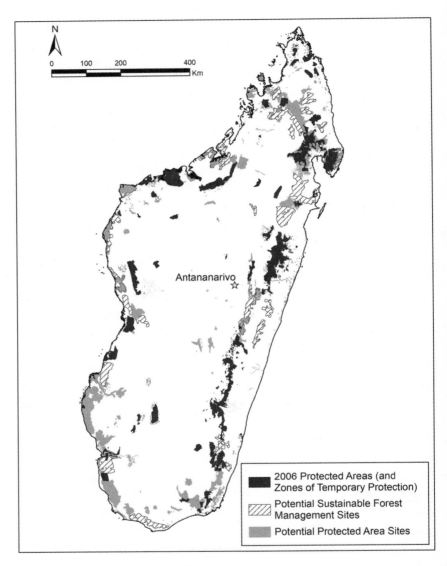

Figure 7.2. Existing and Potential Protected Areas, Zones of Temporary Protection, and Sustainable Forest Management Sites in Madagascar as of 2006. Credit: Catherine Corson. Data sources: The country outline is from ESRI, DeLorme Country data; and the protected areas data is from Rebioma. This data was used in the map accompanying Arrêté Interministériel n° 17914/2006/ MinEnvEF/MEM. Available at http://www.rebioma.net/ [Last access June 6, 2014].

to create them had been lost. COBA members frequently did not have copies of their own maps, and community use patterns bore little resemblance to the delimitations of these maps (Vokatry ny Ala 2006). Similarly, many sacred forests—forests protected by villagers over decades for cultural or religious reasons—were not included on national SAPM maps in part because they were not officially recognized by the state. As a conservation NGO representative succinctly stated, "According to the law, even the sacred forests still belong to the state. . . . People know that it belongs to them traditionally. They are interested in being able to legalize that. But there is no means to legalize that except through the CBNRM projects . . . [with] three- and ten-year management transfers."[25]

Debating Rights to Use Forest Resources

As new borders were being gazetted, government agencies, foreign aid donors, and conservation NGOs based in Antananarivo engaged in lengthy debates over the extent to which the rural people who lived in and around the new parks should be able to use resources within them. Most SAPM commission members agreed that villagers living in and around the parks should not be allowed to exploit forests to the detriment of the biodiversity contained therein, but many also thought that local residents needed economic incentives to preserve forests. The most meaningful debates focused on where and the degree to which use should be allowed and what exactly those economic incentives should be.

While it is difficult to categorize the complexity of various actors' views here, they generally fell into two groups. Relatively strict conservation advocates, typified by CI and WCS, urged that rights be limited to traditional, noncommercial use. They asserted that if communities were permitted to engage in economic activities, they would not be able to stand up to the economic and political power of timber and mining interests and the resulting commercial extraction.[26] These organizations advocated "biodiversity valuation" and nonconsumptive uses, in which sales of medicinal plants, ecotourism, carbon credits, and payments for environmental services would compensate villagers directly for the reduction in their ability to meet their livelihood needs because of park restrictions. They pushed for strict *sites de conservation,* or conservation sites, which would forbid any other form of commercial exploitation within park boundaries.

The countering view was represented by a coalition comprised of German Cooperation, French Cooperation, SAGE, UNESCO, UNDP, and WWF-Madagascar. This group argued that small-scale, as opposed to industrial, commercial extraction would motivate peasants to protect forests. Its views are apparent in the arguments of an American contractor: "If we just restrict forest use, then the forest won't have value for people."[27] The group asserted that sites could operate under strict management plans that the DGEF, NGOs, or other organizations providing technical assistance could approve. Specifically, the DGEF and French Cooperation put forth a concept called Territoire de Conservation et Développement (DCT) (Development and Conservation Territory), which advocated core sites protected for biodiversity and surrounding areas for use that coincided ideologically with IUCN category five (Pollini 2011).

MinEnvEF officials mediated these groups, trying to preserve the financial backing of primarily U.S.-based conservationists while pushing to allow community resource use and protect domestic economic mining and timber interests (Pollini 2007, 2011). Many regional Madagascar government officials as well as regionally based conservation NGOs and contractors were more open to the idea of allowing sustainable use of resources in the new parks than their Antananarivo counterparts. They argued that rural development and agricultural assistance to compensate small farmers for reductions in access to land and resources should accompany SAPM. Yet government officials across a number of ministries and quasi-government departments expressed frustration at the limited funds for development and the overwhelming emphasis by the donor community, particularly the Americans, on conservation issues. One official complained, "In the meetings, it is repeated that we first have to solve the primary problems of food security before thinking of conservation ... yet the donors worry first about conservation, citing Madagascar as a global heritage."[28]

From 2004 to 2006 Antananarivo-based government agencies, foreign aid donors, and conservation NGOs engaged in lengthy debates over the extent to which people who lived in and around the new parks should be able to use resources within them. According to a number of interviewees, visits by IUCN consultants to Madagascar in March and July 2005 were critical in breaking the stalemate over establishing allowable use rights. The consultants emphasized IUCN guidance on protecting human rights, reducing negative impacts on local livelihoods, and including local communities in the management of parks (Borrini-Feyerabend and Dudley 2005a, b). A Madagascar government official commented, "[The IUCN visit] enabled us to unblock a number of things. For example NGOs like CI didn't want to hear of sustainable use. It was an endless debate."[29] SAPM Commission leaders took advantage of the visits to change the name of the new protected areas from "conservation sites" to a "system of protected areas" in an effort to underscore that strict conservation would not prevail,[30] and the DCT concept was superseded by the new SAPM model (Pollini 2011). Despite the initiative's new name, however, advocates for stricter conservation sites remained key leaders in implementing the initiative, and, as one interviewee pointed out, the broad range of IUCN categories permitted everyone to defend their positions using one of the categories.[31]

Fundamentally, the decision to use the Code de Gestion des Aires Protégées (COAP) (Code for Managing Protected Areas) law to implement SAPM bolstered the strict conservationist agenda. As one conservation NGO representative admitted, "There [was] a certain contradiction between what the COAP says and what we plan[ned] to do in the new protected areas."[32] Originally passed as a law in 2001 with backing from USAID, the COAP established the legal framework for ANGAP and the management of natural reserves, national parks, and special reserves. It declared all parks to be the property of the state; it did not establish a legal precedent for community or private management; and it prohibited residing in parks without written permission. Together with associated decrees, it also prohibited all human uses except traditional noncommercial use within park boundaries, and it explicitly forbade commercial exploitation unless authorized as an exception by an individual park management plan. Finally, it detailed various acts that constituted minor and major crimes, punishable by exceptionally strict penalties, again unless the actions were specifically authorized by management authorities. At their most extreme these penalties included twenty years' hard labor, a fine of one billion Malagasy francs (approximately US$100,000), and two years' imprisonment (Repoblikan'i Madagasikara 2001; Repoblikan'i Madagasikara 2005b, c).

In utilizing the COAP as the legal authority for the new protected areas, SAPM introduced the possibility of rescinding the authority to generate economic benefits through forest product use, such as small-scale commercial timber extraction, which had already been granted in CBNRM projects located within protected area boundaries. Both GELOSE and GCF transferred resource management authority for three years, with a potential renewal for an additional ten years (and continual renewals thereafter), pending forest service approval (Hockley and Andriamarovololona 2007). As a result, the management rights granted through these CBNRM programs were fragile, being subject to continued approval by government officials who were advised by the technical aid organizations that facilitated these projects.

Although the COAP law clearly contradicted SAPM's new and improved approach to protected area management by focusing on state management and forbidding commercial resource use, the minister for environment, water, and forests did not want to introduce new parks legislation, having just fought in the national assembly and senate to pass the COAP two years earlier.[33] Instead, the ministry issued two *décrets d'application* (application decrees) to address some of the COAP's limitations. The first decree, in January 2005, stated that protected areas could be managed by entities other than the state and laid out the process such organizations had to go through to create a new protected area (Repoblikan'i Madagasikara 2005b). The second, published in December 2005, defined four new categories of protected areas in accordance with the IUCN categories, including Nature Parks, Natural Monuments, Harmonious Protected Landscapes, and Natural Resource Reserves; it stressed that for category five, extractive activities must be traditional and that at least two-thirds of the area of any category six park should be in its natural state (Gardner 2011; Repoblikan'i Madagasikara 2005c). As Pollini (2011) notes, the decree did not require community-level consultation or compensation or a socioeconomic impact assessment. Furthermore, neither decree specifically challenged the COAP's strict restrictions and penalties, and both underscored that individual park managers had the authority to determine acceptable uses. This legal guidance left the final decisions about allowable resource uses within the new protected areas up to the entities putting together individual management plans—who were likely to be, under the new governance regime, conservation NGOs or private businesses. In this manner they strengthened the power of nonlocal and nonstate agents to determine acceptable local resource uses and undermined preexisting community management systems.

The Rush to Implement the Durban Vision

Numerous government and donor policies highlighted the importance of consulting with potentially affected communities. IUCN consultants stressed the need to engage regional actors and mayors in the planning process (Borrini-Feyerabend and Dudley 2005a). SAPM guidelines, drafted collaboratively by NGOs, government officials, and donor representatives, emphasized that the creation of new parks should involve consultation with affected populations and promote community management (Commission SAPM 2006). Legislation embodied in a decree of December 2005 required consultation with people having customary and state-recognized land rights (Repoblikan'i Madagasikara 2005c). Finally, any protected areas funded by the World Bank were compelled to comply with environmental and social safeguard policies, which stipulate that bank projects avoid or limit any involuntary resettlement or substantial loss of livelihood (World Bank 2005).

However, rural residents living in the Ankeniheny–Zahamena and Fandriana–Vondrozo corridors, where I concentrated my study, expressed very little knowledge of SAPM.[34] When I first discussed the program with SAPM Commission members in October 2005, they were rushing to establish 1 million hectares by the end of the year, and SAPM had become the key topic of discussion among donors, government officials, NGO representatives, contractors, and others involved in Madagascar's environmental program. Between the November 2003 announcement and December 2004, commission members focused on determining the priority areas for conservation, and no new parks were established. By January 2005, with only four years left to meet the president's target, the minister for environment, water, and forests decided to establish 1 million hectares of protected areas per year. This target formed the proximate cause of the rush.

By October 2005 only about 100,000 hectares of new protected area had been created. It was clear that the government would not meet its target of 1 million hectares by the end of 2005 if extensive rural consultations about SAPM took place. In particular, it was impossible to consult with all potentially affected rural villages in the two eastern rainforest corridors, which comprised the two largest proposed protected areas. A massive debate ensued in Antananarivo over how much consultation was really necessary, and it split groups who might otherwise have been allies. A number of national and regional Madagascar government staff stressed the importance of thorough consultation so as not to cause resistance and burning of forests as a result, and organizations such as WWF-Madagascar and regionally based U.S.-funded contractors advocated a slowdown in the process. Others argued that, given the importance of Madagascar's biodiversity, it was critical that the program be implemented rapidly, and CI in particular pushed to meet the deadline.

Ultimately, in order to meet the minister's deadline but to allow for future consultation, the SAPM Commission decided to establish new parks under orders that granted temporary protected status to the new protected areas that was renewable after two years and that delineated initial park boundaries and allowable resource uses based on limited consultations. These orders left specific details about park management and uses within subzones of larger parks to be decided in future management plans. Moreover, given that the final policy guidance and legislation were not issued until 2008 and 2009, these temporary parks were established without clear guidance on how to develop management plans or conduct community consultations. Limited community engagement took place at the mayoral level, but the rush to meet the five-year deadline superseded consultation efforts with villagers themselves, and consultation leaders or Antananarivo-based SAPM leaders or both suppressed any mayoral dissent (Corson 2012).

The Rural Consultation Process

The organizations involved in the community consultation process that did occur reflected the makeup of the Antananarivo-based SAPM Commission. Regional representatives of the DGEF and ANGAP as well as primarily USAID-funded conservation and agricultural development NGOs and contractors formed committees to design and implement the SAPM initiative. The USAID-sponsored Miaro conservation program, administered by CI in collaboration with ANGAP, WWF-Madagascar, and WCS, provided much of the funding and technical guidance. DGEF and CI representatives based in the regional cities of Toamasina, Moramanga, and Fianarantsoa coordinated the initial consultations in the two eastern rainforest corridors in 2005. In 2006 the DGEF took on a larger coordination role, although without funding or practical implementation capacity: bureaucrats often worked without phones, faxes, computers, and vehicles. Moreover, because American grantees and contractors could more easily communicate with their Antananarivo-based counterparts, they often received information about SAPM Commission decisions before their Madagascar government colleagues did. Thus they continued to play a behind-the-scenes role, facilitated by their human resources, vehicles, and GIS skills.

The process of delimiting the new protected areas in these corridors began in both regions with biodiversity prioritization exercises that drew on aforementioned scientific endeavors to expand conservation planning to a landscape scale, such as the GEF workshop of 1995 (Hannah et al. 1998), the USAID-funded National Protected Area Management Plan, or PlanGrap (ANGAP 2003), and the REBIOMA initiative (Kremen et al. 2008; Randrianandianina et al. 2003). Although none of these efforts had incorporated socioeconomic data, such as what resources local residents depended on for their livelihoods, the resulting proposed boundaries for the new parks were used to negotiate with community leaders during rural consultations. Held at the district level, the supposed consultations aimed to convince the population to agree to the proposed boundaries. A conservation NGO representative explained the process: "In the beginning of the year [2005] we organized a ... scientific workshop ... mostly based on biodiversity data, and we identified which part of the corridor we should preserve for this or that species . . . this map will be the grounds for negotiation with the communes."[35]

In both corridors pressure to reach the minister's annual deadline, combined with lack of staff and funding for either government agencies or NGOs to reach villages located in remote areas, meant that consultation efforts reached only the district level, relying on mayors as the spokespeople for the tens of thousands of people living in the communes. One member explained that, because it would be more than a two-day walk to reach certain sites, it would require human resources, time, and material they did not have to conduct a thorough consultation. In the Fandriana–Vondrozo corridor, a regionally based NGO representative in Fianarantsoa said, "We tried to gather people because it was impossible for us to visit sixty-six municipalities with ten districts and five regions."[36]

In theory villagers could later contest the park boundaries agreed upon during these discussions. But in order to grant the temporary protected status establishing the new parks, key political figures, including ministers, regional political leaders, and mayors, had to agree to the initial limits. Thus by the time the temporary protection status was granted there was already a substantial buy-in at various political levels. Furthermore, such an approach empowered mayors as the local decision makers and the primary liaisons between national and regional governments and residents. It assumed that mayors would consult with the affected populations and ignored intra- and intervillage tensions, gender and class differences, and mayors' tendencies to favor their own villages (Kull 2002a). Several interviewees reported that mayors did not consult with villagers before agreeing to particular limits.[37] One interviewee recalled the top-down process of deciding which areas should be included: "I attended a meeting in Fianarantsoa about SAPM. The mayor and the president of COBA attended too. Then the SAPM person said, 'Set boundaries for the areas I list.' Then the mayor set boundaries. Then [the SAPM team] wrote a report that said, 'Our mayor already signed this.' That's what happened."[38]

Sustaining this assertion, a 2006 survey that the Malagasy NGO Vokatry ny Ala helped me to conduct on village awareness of SAPM among preexisting CBNRM sites in the communes of Ikongo, Ambatofotsy, Miarinarivo, Andranomiditra, and Ialamarina in the Fandriana–Vondrozo corridor found that, with the exception of mayors and presidents of COBAs, very few individuals had heard of SAPM. The mayors and COBA presidents had attended SAPM consultations but in most cases had not reported back to villagers or solicited their input. One COBA president justified this way of handling the consultations by saying he felt it would be better to report to the population after the final policy had been decided. Of the 130 people interviewed, 75 percent had never heard of SAPM, and 85 percent of the 32 people who had heard of it were mayors, COBA presidents, or COBA members who had been invited to SAPM sessions (one commune held a workshop to which it invited some COBA members). The other five had heard about it on the radio. Villagers who knew about SAPM thought it was a form of CBNRM, and even COBA presidents and mayors saw it as a mechanism that would help secure local access and control over resources by officially recognizing, albeit not officially transferring, land rights (Vokatry ny Ala 2006).

Intriguingly, when talking about the consultation process, interviewees often conflated consultation with *sensibilisation*, a French term which can be translated into English as "persuasive education," "outreach," or "awareness-raising." Development organizations often use the term to refer to their field training activities, such as health education and agricultural improvement techniques. For example, a Fianarantsoa-based government official said, "In the districts we conduct the *sensibilisation*-consultation. We say that to implement the president's declaration, you need to give a part [of your land] to be categorized according to IUCN guidance."[39] Similarly, a Toamasina-based official noted, "We tried to inform the participants about the necessity of conservation and the important role played by forest resources."[40] Another Antananarivo-based official commented, "People should understand what [SAPM] is, and that it is for their own good."[41] Thus the goal was not to consult the population about whether or not they wanted the parks, but to instill in them the notion that parks were for their own good. Consultation began with informing the participants what they *should* think, specifically, that the conservation areas were necessary. Effectively, the processes in the two corridors served to convince the population to agree to predetermined areas based on biodiversity priorities.

I contend that regional consultation leaders conflated consultation with sensibilisation because their mandate from Antananarivo was to persuade, not to consult. While various workshops designed to raise awareness about the initiative were held throughout the country, there was no clear guidance on how to conduct local consultation. The decree of December 2005 simply said that affected populations should be consulted and their interests taken into account but did not say how to do that (Repoblikan'i Madagasikara 2005c). Similarly, IUCN consultants stressed the need to "engage at least some community representatives in the preliminary identification of the protected areas," followed by more extensive consultation later (Borrini-Feyerabend and Dudley 2005a, 14). However, their guidance on how to conduct consultation stated merely that it should occur through "proven and coherent methods and tools, in particular with regard to social communication and participatory governance approaches" (Borrini-Feyerabend and Dudley 2005b, 11). Finally, draft policy guidance said to "begin consultations with communal councils and / or mayors, regional authorities, technical services, and development programmes to ensure their commitment towards the creation of new areas protected" (Commission SAPM 2006), and to consult with communes, villages, and hamlets only before final park creation. While it did not define consultation or how the interests of stakeholders should be taken into account, it did underscore that the goal was to ensure local commitment to park creation (Commission SAPM 2006).

Regional representatives of the Madagascar government, NGOs, and contractors had to mediate between conservation visions emanating from Antananarivo and local realities. While the Antananarivo-based commission members debated how much consultation would be necessary, regional staff had to carry out their mandates. Regional agents complained about disorganization, incoherence, and political pressure from Antananarivo-based policy makers who seemed unaware of the challenges they faced in implementing the program. One regional government official said, "CI and other donors, they all have different concepts . . . so we are lost. Yet at the same time they tell us to do it all before 2005."[42] In particular, regional contractors, government officials, NGO staff, and mayors repeatedly emphasized that SAPM should be accompanied by rural development assistance to compensate villagers for reductions in access to land and resources. Yet very limited funds materialized to this end, and Bertrand et al. (2014) also notes a decline of funding for accompanying sustainable development that has impacted communities.

Finally, interviewees reported that mayoral resistance was stifled. In the Ankeniheny–Zahamena corridor, where CI was avidly pushing for the rapid establishment of a protected area, a regional contractor commented, "The tendency was that commune mayors didn't want to make definitive decisions. But that got lost in the shuffle, perhaps due to the bias of the people getting pressured to hurry."[43] In the Fandriana–Vondrozo corridor a USAID contractor more openly discussed mayoral resistance: "We still see much reticence from people, the COBA, [and] the mayors [because] the protected areas with which they are familiar are those managed by ANGAP. . . . So when we say we are going to increase the extent of protected areas . . . [they think] it will constrain their way of life."[44] During the initial mayoral discussions, the representatives of the government, NGOs, and contractors who led the consultations did not use the biodiversity prioritization map because it did not have communes marked on it, and mayors needed to see political as well as ecological boundaries. Instead, in an effort to introduce some bottom-up decision making, the leaders asked mayors to propose areas within their communes for protected status. The resulting map contained a series of unconnected parks that failed to meet the fundamental conservation goal, which was to preserve the forest corridor. The central SAPM technical secretariat instructed the consultation team to go back in order to *persuade* the mayors to create a connected corridor, which they did.[45] The delimitations from this amended mayoral consultation were then included in the 2006 1 million hectares of protected areas.

In spite of these shortcomings the two major eastern rainforest corridors, together comprising almost 1 million hectares, were designated as temporary national parks in 2005 and 2006. The Ankeniheny–Zahamena corridor was included in the 919,000 hectares of protected areas established in 2005 (Repoblikan'i Madagasikara 2005a), and, because of the mayoral dissent, the Fandriana–Vondrozo corridor was delayed until 2006 (Repoblikan'i Madagasikara 2006b).

At the end of 2006 DGEF agents in both regions stressed that they hoped to do village-level consultation in the future. However, they had neither the funds to pay staff to moderate the discussions or to cover vehicle expenses to get to remote locations, nor the equipment or staff to develop maps. So they continued to rely on NGOs and consultants.[46] The CI regional office, however, did not have a budget for such consultations, and the disbursement of World Bank funding for consultations from the national DGEF office to the regional offices was delayed for various bureaucratic reasons. A central topic of debate during the third EP3 supervision mission in 2006 was how to compensate villagers for the impacts from the new protected areas, given the limited funding for development activities and the World Bank's social and environmental safeguards. In the end, a national framework for the mitigation of, as opposed to compensation for, these impacts was developed (Pollini 2011). So, despite the SAPM leadership's insistence on the importance of consultation and comanagement of parks and the millions of dollars of foreign aid for environmental issues in Madagascar, little money materialized for village consultations or compensation, revealing that, contrary to the participatory rhetoric, participation was not a donor or an NGO priority.[47]

Protected Area Management

In 2008 and again in 2010 the government renewed the 2006 order with an updated map of protected areas that included priority sites in addition to potential sites (fig. 7.3); introduced a revised COAP law; and issued a series of policies that both allowed limited commercial resource extraction and reinforced nonstate powers. Policy documents issued in 2008 and 2009 detailed the conditions under which commercial exploitation could be allowed in SAPM's more flexible parks and specifically in CBNRM sites. They also mandated that management plans be prepared in a "participatory" manner, include provisions on community rights, and be informed by socioeconomic studies. Finally, they underscored the need to compensate villagers for any reductions in resource access caused by park establishment (Repoblikan'i Madagasikara 2008a, b, c, 2009a, b). Additional ethnographic research would

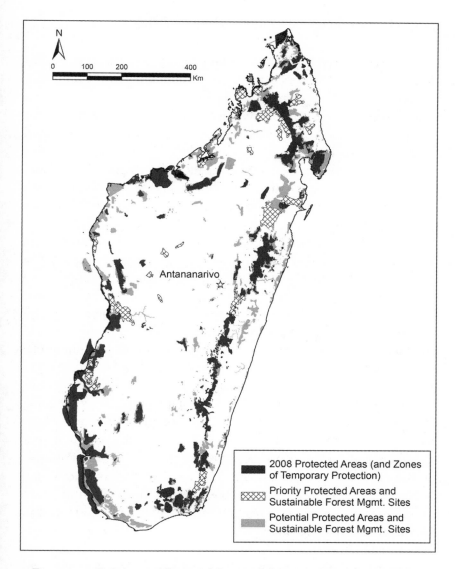

Figure 7.3. Existing and Potential Protected Areas, Priority Protected Areas, and Sustainable Forest Management Sites in Madagascar as of 2008. Credit: Catherine Corson. Data sources: The country outline is from ESRI, DeLorme Country data, and the protected areas data is from Rebioma. This data was used in the map accompanying Arrêté Interministériel n° w18633/2008/MEFT/ MEM. Available at http://www.rebioma.net/ [Last access June 6, 2014].

be needed to assess the politics behind the documents and their impact.

In December 2010 the transition government passed two additional orders that renewed the 2008 temporary protection order, streamlined some of the regulations, included more explicit restrictions on the use of resources in marine and coastal protected areas, created an intergovernmental SAPM Commission, and codified 171 new protected areas and sustainable forest management sites covering 8.7 million hectares in total (fig. 7.4) (Repoblikan'i Madagasikara 2010a; Repoblikan'i Madagasikara 2010b). Yet at the end of 2013, of the 94 temporary protected areas, only 8 had permanent status (Bertrand et al. 2014).

As much as these changes promised improved consultation, several problems remained. The 2008 guidelines reiterated that a goal of consultation was to "convince the surrounding population of the benefits of sustainable resource management and the protected area" (Repoblikan'i Madagasikara 2009a, 17), and they still required only commune- or mayoral-level consultation for the preliminary establishment of the park. In addition, while commercial activities were allowed in some parks, they had to be approved in individual park management plans, which were developed by individual park managers—who could be state agencies, sponsoring donors, conservation NGOs, or private companies. Hence villagers' ability to conduct both livelihood and commercial activities still depended on the continued support of these organizations as well as on the relative enforcement of the COAP versus local law, as Keller (2008) discusses. Whether or not the commitment to consultation will more effectively translate rhetoric into practice in the future will depend on overcoming the demonstrated inadequacies of earlier efforts.

Regardless, nonstate actors have continued to influence land tenure politics, particularly in the ongoing negotiation of mining rights in the new protected areas and in the rapid translation of global targets into paper parks. A World Bank document from 2011 highlights the persistently strong influence of transnational NGOs on Madagascar conservation policy as they successfully convinced the transition government to further expand protected areas immediately after the CBD Conference of Parties (CoP)

Figure 7.4. Existing and Potential Protected Areas, Priority Protected Areas, and Sustainable Forest Management Sites in Madagascar as of 2010. Credit: Catherine Corson. Data sources: The country outline is from ESRI, DeLorme Country data, and the protected areas data is from Rebioma. This data was used in the map accompanying Arrêté Interministériel n° 52005/2010. Available at http://www.rebioma.net/ [Last access June 6, 2014].

decision of 2010 to increase the protected area target from 10 percent to 17 percent of a country's terrestrial area—an increase that CI had strongly lobbied for at the meeting (Campbell, Hagerman, and Gray 2014; Corson et al. 2014). Using numbers that exclude 2.5 million hectares of sustainable forest management sites, the report summarizes as follows:

> The protected area network in Madagascar [includes] 2.4 million hectares of protected areas managed by Madagascar National Parks and 4.5 million hectares of new protected areas that are being developed predominantly by NGOs (including CI, WCS and WWF). . . . Triggered by the Convention on Biodiversity Conference of Parties (CBD-CoP) in Nagoya in October 2010, informal discussions have recently commenced amongst the Government and NGOs as to the feasibility of increasing coverage of the network to cover 16–18 percent of the country's surface. (World Bank 2011a, 31)[48]

Reinforcing Upward Accountability in Conservation Governance

Despite SAPM advocates' insistence that the creation of new protected areas would engage communities that live in or utilize resources in the parks, minimal village-level consultation took place in the two eastern rainforest corridors studied. Notwithstanding the millions of dollars for conservation in Madagascar, Antananarivo-based donors, government agencies, and NGOs provided inadequate finances for and guidance on how to conduct local consultations. Furthermore, they debated what kinds of resource uses they thought should be allowed and what the economic incentives for conservation should be in the absence of information from villagers about what they needed and wanted from the forest. National and international political pressure to implement the program rapidly required that the regional staff's day-to-day attention be focused not on consultation with affected villagers but on creating biodiversity prioritization maps and meeting numerical targets. This pressure led to a consultation process that reached only the mayoral level, where mayors—empowered as liaisons between the national government and villagers—often failed to disseminate information to villagers, who consequently remained unaware of the program. Additionally, the consultations comprised processes of persuasive education that began with maps that outlined park boundaries based on biodiversity priorities rather than on local resource use patterns, and in which mayoral resistance was stifled in order to reach annual targets. Regional consultation leaders conflated consultation with sensibilisation because they were directed to persuade, as evidenced by the push to establish parks quickly, the request to redo the mayors' map of unconnected sites, and the guidance to ensure local commitment to conservation. Ultimately, the initiative superseded, rather than supported, previous community conservation initiatives by introducing the possibility of rescinding previously granted community commercial use rights (Corson 2012). Through these various processes, despite claims of recognizing local rights, SAPM legitimated the authority of consultants, private companies, scientists, transnational conservation NGOs, and foreign aid donors to shape Madagascar's future forest policy, including who could benefit financially from forests. Even as it was shrouded in rhetoric of community participation, in practice it reintroduced top-down planning, thereby reversing the decentralization initiatives started in the 1990s and reinvigorating preservations and back-to-barriers approaches (Bertrand, Horning, and Montagne 2009; Blanc-Pamard, Pinton, and Rakoto Ramiarantsoa 2012).

The Madagascar case reveals not just the potent influence of transnational NGOs and private sector interests on domestic forest policy but also how state processes can be produced by unelected and unaccountable entities, dispersing both claims to and authority over natural resources and subjects across local, national, and international boundaries. By participating in official committees, ranging from the Malagasy delegation to the parks congress to the SAPM implementing committees, foreign and nonstate actors also cemented their power and authority to determine the validity of particular claims to forestland. In this manner the state gave way to the consortium of state officials, foreign donors, transnational NGOs, contractors, and private sector agents. In short, SAPM not only privileged the claims of the forest service, foreign conservation organizations, and mining interests over those of rural peasants but also legitimated the power and authority of these agents rather than those of rural peasants to determine forest policy, to control the source of rural livelihoods, and to garner wealth from forest resources.

CHAPTER EIGHT

Transforming Relations of Power in Conservation Governance

Marc Ravalomanana, the president in office from 2002 to 2008, helped Madagascar make the greatest progress in conservation in its history as a country. . . . His conservation vision did not come out of the blue. It was supported by a number of major international conservation organizations—notably WWF and CI—as well as national NGOs, the World Bank and a twenty-five-year effort by USAID, which I believe is the best biodiversity conservation effort ever supported by USAID anywhere in the world.

—RUSSELL MITTERMEIER, former president of Conservation International, March 3, 2014

There has been notable progress in all areas where USAID has worked: at the policy and institutional level, in park management and the creation of protected areas, in reducing

pressures on forest resources in particular locales. These
successes have been much celebrated and have kept hope
alive that indeed progress is possible. Yet, as we step back
after 25 years, honesty compels us to conclude that the
environmental crisis in Madagascar is far more acute now
than it was at the outset of EP1.

—KAREN FREUDENBERGER, in *Paradise Lost? Lessons from
25 Years of USAID Environment Programs in Madagascar*

THE JUXTAPOSITION OF THESE quotes by two well-known, respected
foreign actors in Madagascar environmental politics prompts us to
ask why, if USAID's biodiversity conservation program is the best
conservation program ever supported by USAID, the environmental
crisis in Madagascar is more acute than it was twenty-five years ago?[1]
For one, USAID's interventions are only one of many influences.
Numerous factors that have shaped degradation in Madagascar are
beyond USAID's control. Most important, if we recall Mosse's
(2005) proposition that the success of a development project is produced not through its impacts in the field but through its ability to
stabilize social relations, we can see that the statements are not mutually exclusive. USAID's conservation program mobilized and sustained relations among a diversity of actors, including conservation
NGOs, other bilateral and multilateral donors, mining companies,
Madagascar government agencies, the U.S. Congress, the U.S.
Department of State, contractors, and scientists. Yet the combined
consolidation of state politics within global environmental institutions and the introduction of neoliberal reforms that together liberalized economies and opened up opportunities for nonstate actors to
influence state policy processes conditioned the way in which these
actors interacted. The programmatic priorities and associated narratives that emerged reflected dynamic power relations among them,
and the need to maintain these relationships directed funding and
attention toward narratives that attributed blame and responsibility
for Madagascar's environmental crisis in ways that all parties could
endorse. In this manner the compromises needed to maintain these

political coalitions hindered the pursuit of a more comprehensive
and sustainable conservation approach.

Ethnography of Policy Making Across Time and Space

By traveling across policy-making sites from the halls of the U.S.
Congress to donor meetings in Antananarivo to villages in eastern
Madagascar, I have emphasized the intertwined nature of the politics
of U.S. foreign aid and struggles over land and resources in
Madagascar. Through multiple ethnographic windows, I have shown
how the seemingly mundane, everyday practices of individuals inter-
sect in particular ways at particular moments in time to catalyze
transformations in policy processes and priorities. Attention to these
moments reveals foreign aid not as an imposed Western machine
but as a process of compromise and consent, where power is not inherent in structures but relational and dynamic, formed through
everyday practices that transcend geographic and institutional
boundaries.

The growth of USAID's environmental programs and introduction of Madagascar's NEAP at the historical moment of the late
1980s shaped which political strategies and narratives became viable. Preexisting ideas, narratives, and policies conditioned those
that emerged. In the United States the lack of public support for
foreign aid, the rising awareness of a global fuel crisis, and the
introduction in the 1970s of the New Directions legislation
established the basis for environmental advocacy groups to convince Congress to legislate initial USAID environmental policy.
Likewise, foreign scientists interested in Madagascar's biodiversity
created the foundation for the NEAP biodiversity program.
Finally, a long history of blaming slash-and-burn agriculturalists
for degradation while enabling large-scale commercial industries
to deforest fed into contemporary degradation narratives and solutions in Madagascar.

Even as these histories informed subsequent events, specific
conjunctures reshaped the realm of the possible in novel ways. The
rise and articulation of neoliberalism, sustainable development,
participatory development, and biodiversity conservation in the
mid- to late 1980s offered the confluence of political forces needed

for scientific advocacy around Madagascar's biodiversity to gain political momentum. Similarly, the turn to focus on the IFIs by many of the initial environmental advocates who had activated congressional support left a vacuum for the biodiversity-focused NGOs to become the primary political advocates for USAID's environmental programs. Finally, the downsizing of the government in both Madagascar and the United States created opportunities for nonstate actors to become more involved in state policy.

Within these political-economic contexts, particular moments, ranging from a brainstorming session in a conference hotel room to a presidential speech; events such as major conferences; institutions such as the ICCF; and places such as Washington, D.C., served as nodes of connection at which various state and nonstate actors came together to negotiate programs, policies, relationships, and narratives. As they did so, they constructed political arenas for the circulation and negotiation of particular narratives and ideas.

At these nodes, individual actors were able to shift policy trajectories. Actions ranging from drafting conservation plans to writing congressional legislation to strategically using biodiversity funds to support integrated agriculture and environment projects contested and reshaped political forces toward particular objectives. While the biodiversity earmark guidelines hindered USAID mission officials' ability to pursue integrated environment and agriculture projects, mission staff creatively and strategically worked around bureaucratic constraints to link conservation with agricultural intensification, governance, health, and rural development. In this sense there was a dialectical relationship between individual agency and structural power: agency was prefigured but not determined by structural constraints, and power was not inherent in structures but relational and continually formed, maintained, and contested through interaction. As actors came together in these moments they legitimized and shaped disparate pursuits into common policy trajectories, and the shifts in meaning and practice they negotiated became institutionalized in policies, relationships, and programs that comprised a dynamic field of governance.

Shifting Relations of Environmental Governance Under Neoliberalism

These ethnographic windows show how the rise of biodiversity conservation in USAID's environmental portfolio was entwined with the reconfiguration of state, market, and civil society relations under neoliberalism. Whereas environmental and foreign aid policies in the 1970s emphasized state regulation and protection of human welfare in order to reduce the negative environmental effects and unequal benefits of capitalism, by the 1980s the sustainable development discourse had enlisted the environmental movement in the pursuit of economic growth, and the state had given way to privatization, deregulation, and liberalization. USAID opened its mission in Madagascar precisely because the government there had agreed to neoliberal structural reforms, and its continued presence was dependent on adherence to structural adjustment. These reforms advocated rapid economic growth through reduced state regulation, increased foreign direct investment, and liberalized trade, which collectively created incentives to increase foreign investment in and reduce government regulation of natural resource exploitation.

The implementation of these political and economic reforms changed who got to sit at the political negotiating table. Policies from the Reagan to Clinton administrations that downsized USAID and contracted out services to private and nonprofit actors contributed to the rise of these actors in development politics and conditioned their eventual ability to shift public investments toward their particular interests. In this process NGOs, despite the corporate partnerships that increasingly pervaded their operations, capitalized on post–Washington Consensus idealized visions of civil society to transition into fundamental components of the political infrastructure tasked with environmental governance.

The reconfiguration of public, private, and nonprofit relations also changed what was politically tenable to do. As the assemblage of NGOs advocating for USAID's environmental program evolved from the initial group of environmental activists pushing a broad-based environmental agenda to a coalition of USAID grantees with a specific programmatic interest in biodiversity conservation, the focus of USAID's environment program narrowed around the interests of its political champions. Likewise, the World Bank's push to engage private and nonprofit actors in the development of priorities for Madagascar's NEAP led to a diverse environmental agenda, but one that was constrained by available funding and the policies entailed in structural adjustment. The Madagascar NEAP and its associated Environmental Charter as well as subsequent overarching policy documents reflected political attempts to balance a reliance on natural resource–driven economic growth with conservation priorities. They emphasized individual responsibility for change, particularly by rural peasants, the reduction of state regulation, the promotion of natural resource–led growth, and the expansion of protected areas.

Finally, the reconfiguration led to a mutual dependence among state and nonstate actors. As both the U.S. and Madagascar states turned to private and nonprofit organizations to implement their work, they became dependent on these partners not just to design, implement, and evaluate environmental programs but also to foster political support for them. USAID funded conservation NGOs to undertake projects and offered political legitimacy to their position as negotiating partners in policy arenas. In turn, the Washington-based biodiversity lobby secured congressional backing not just for environmental funds but also for the USAID-Madagascar mission. The mission's overt prioritization of biodiversity and emphasis on national parks were successful strategic maneuvers in maintaining this support in Washington. For newly interested members of the U.S. Congress, international conservation represented an easy way to support environmental issues without angering their key constituents or confronting other members. For the Madagascar government, the endorsement of biodiversity conservation cemented relations with critical Washington-based foreign aid sources.

The resulting horizontal partnerships among these actors centralized decision making, the opposite of what one might presuppose. As these actors invested resources and personnel in coordination among themselves, they concentrated decision-making authority in capital cities, where policy makers were accountable not to those affected by their policies but to those far removed from the effects of their decisions. With respect to SAPM, the immediate demands of the president and the minister of environment, water, and forests drew attention away from community conservation and consultation with rural peasants. Instead, Antananarivo-based organizations focused energy and funds toward satisfying funders, meeting global targets, and building political constituencies in Washington rather than engaging in rural consultations. Likewise, in utilizing the biodiversity earmark, the USAID mission was beholden to members of Congress and Washington-based foreign policy makers, who determined what qualified for biodiversity funds. In short, while the high-profile nature of Madagascar's biodiversity sustained donor investment in conservation, it also encouraged upward rather than downward accountability all along the chain—from the U.S. Congress to Malagasy villages.

Isolating the Environment "Over There"

By framing conservation as a peasant problem that took place "over there" advocates sustained environmental foreign aid investments in Madagascar. They engaged diverse interests, from conservation NGOs to the international financial institutions to mining companies, by offering biodiversity conservation as a political means of being green without taking up more difficult environmental issues. For many members of Congress international conservation was a way to be an environmentalist without having to endorse controversial environmental issues like climate change. For the Madagascar government it could be pursued without challenging neoliberal reforms, including economic growth based on expanding resource extraction. For the conservation NGOs, it was a means to continue advocating for the conservation of biodiversity by separating nature from humans—an ideology that, even as it is debated, remains at the core of the conservation movement. These actors transformed biodiversity into a high-profile issue within development politics by isolating the environment as something that could be set aside in a park over there, in foreign or peasant lands or both, and distant from competing economic and political interests. Nonetheless, this approach reinforced the separation of environmental issues from mainstream policy, allowing for the conceptualization of conservation without changes in distribution of economic or political power. It offered a green stamp of approval that enabled companies to "offset" their extraction and the environmental degradation caused by their practices by supporting the creation of parks elsewhere, allowed otherwise antienvironmental politicians to become green, and gave Madagascar government officials a way of satisfying environmental donors while also pursuing economic growth based on increasing resource extraction.

The lack of viable management capacity for Madagascar's future protected areas did not deter the program's advocates precisely because the program provided a means of maintaining a cohesive strategic transnational alliance behind Madagascar's conservation program. For the NGOs it created a focus for fund-raising campaigns and cemented political access toward conservation goals. For USAID it was a way to demonstrate progress in meeting the biodiversity earmark. And for the Madagascar government it attracted additional foreign aid.

What Is at Stake for Madagascar?

The impact is that, in spite of a well-organized and well-funded environmental program, Madagascar still lacks the institutional capacity to control the expanding commercial exploitation of its natural resources. While shifting cultivators are deforesting Madagascar's eastern rainforests, they are acting within a broader political economy that has pushed them onto marginal lands and secured profitable lands for large-scale commercial enterprises. The focus on changing the practices of rural people and rapidly expanding protected areas has directed attention away from this broader context as well as other critical environmental issues, from urban pollution to education infrastructure to sustainable trade. It has undermined the conservation agenda itself, leaving the country with a network of paper parks, alienated villagers, and insufficient state capacity to contest corruption, collusion, and exploitation of timber and minerals within parks.

In 2009 Andry Rajoelina overthrew Marc Ravalomanana in an army-backed coup. Previously the mayor of Antananarivo, Rajoelina campaigned in part as an advocate for the people disenfranchised by Ravalomanana's procapitalist policies and close business ties. Elected in 2002, Ravalomanana had simultaneously embraced neoliberalism and become a darling of the conservation community. But among the mistakes the former president was accused of committing was negotiating with a South Korean company to lease 1.3 million hectares of land to farm maize and palm oil. As a deal perceived to threaten rural land tenure security, it is one example of how concern about international control over Madagascar's natural resources contributed to the overthrow of a leader widely acclaimed in conservation circles as Madagascar's first environmental president.

Many foreign donors suspended their aid to Madagascar in protest of the coup, and organizations such as the African Union and the Southern African Development Community banned Rajoelina's transitional government. Nonetheless, foreign NGOs continued to support environmental programs in Madagascar and successfully extracted commitments for further protected area expansion from the transitional government. A new constitution was passed in 2010, elections finally took place in January of 2014, and Hery Rajaonarimampianina, a cabinet member during Rajoelina's transitional government, became the president of the Fourth Republic. As the elections were deemed to be free and fair and the African Union welcomed Madagascar back, foreign aid began flowing again.

At the Sixth WPC in Sydney, Australia, in 2014 President Rajaonarimampianina reiterated Madagascar's commitment to the Durban Vision, promised to ensure effective management of existing protected areas by 2015, and agreed to triple marine protected areas by 2020. He also vowed to address the worsening illegal trade in rosewood, palisander, and ebony, which exploded in the aftermath of Rajoelina's coup. USAID has similarly indicated that it intends to "assist community-based organizations in monitoring illegal activities involving Madagascar's endangered natural resources and compliance with laws; encourage local organizations to fight these illegal practices; and make the media more aware of them" (USAID 2014b). While these are admirable ambitions, their translation into effective and equitable conservation will require transforming the power relations that have both created Madagascar's environmental crisis and shaped previous interventions to redress it.

This transformation must begin by questioning the goal of expanding protected areas without ensuring an effective management regime first. Even before SAPM the country's main challenge was not to create more protected areas but to effectively manage those it already has. The initiative to triple Madagascar's protected areas took as a starting premise that Madagascar had 1.7 million hectares of existing parks, which included only the national parks, strict nature reserves, and special reserves that ANGAP managed for conservation. However, utilizing the IUCN's progressive protected area categories, which allow sustainable use, Madagascar had met the 10 percent goal before the SAPM announcement. As Simsik (2003) notes, if biosphere reserves, hunting reserves, forest stations, and reforested areas as well as the 4 million hectares of classified forests that the forest service managed for wood supplies are included, a total of 6.6 million hectares were already protected in Madagascar, on paper if not in practice, by the early twenty-first century.

At best, the new parks are destined to be paper parks. As a regional contractor challenged, "With no way to enforce the management objectives what happens is business as usual."[2] The lack of government capacity to manage forests owing to mutually reinforcing deficits of human, material, and financial resources was well documented before SAPM. At the conclusion of this research, ANGAP, renamed Madagascar National Parks, still did not have the capacity to manage the parks, and even as efforts were being made to raise funds though the Madagascar Biodiversity Fund, most parks remained reliant on foreign aid to fund recurrent costs.[3] In the best-case scenario, tourism could generate only 10–15 percent of park operating costs (Méral et al. 2009). Second, even if it had financial sustainability, Madagascar National Parks would need legal authority to enforce park regulations, as only the DGEF could sanction violations within protected area borders (Montagne and Bertrand 2006).[4] In 2012 there were roughly 294 forest officers and 6 million hectares of protected areas, which translates to one agent per 20,000 acres (Brimont and Bidaud 2014). While this may be the most promising outcome for resident communities who, in the absence of external enforcement, could continue to utilize resources, more powerful actors could also access resources. Even where environmental impact assessments were required for mining exploitation, the government did not have the capability to enforce the requirement. Mining permits had been granted throughout protected areas, and most illegal timber extraction took place in protected areas.

At worst, the creation of new protected areas will foster *increased* deforestation: many authors have argued that SAPM has already contributed to a rush to capture land by clearing it (e.g., BATS 2008; Freudenberger 2008; Pollini 2011). As contemporary conservation investments reproduce historical conflicts over land tenure, resource rights and authority, and illegal and legal commercial extraction, they threaten to contribute to, rather than address, the degradation of Madagascar's forests. Simsik (2003) warned more than a decade ago about the potential environmental efforts had for catalyzing increased forest exploitation: "It is possible that IENGO [international environmental NGO] efforts may have actually accelerated the rate of environmental destruction in Madagascar. Once outsiders began showing great interest in resources that were not their own, they inadvertently created a situation whereby local residents felt the need to exploit whatever natural resources they could while they were still available, fearing that they would eventually lose their open access to these resources by having the land placed in some protective status by the state" (128). Many regional governmental and nongovernmental actors echoed this fear; Madagascar government officials in particular expressed concern that the protected area expansion would prompt small farmers to burn forests in protest.[5] A USAID consultant report similarly argued that "among many rural people SAPM gained an early reputation for being top-down and largely engineered by outsiders. This has bred skepticism and hostility that will be difficult to overcome" (Freudenberger 2010, 47).

Transforming Relations of Power in Conservation Governance

Recognizing the basic human rights to resources of local villagers is not just a moral imperative but also a practical one, as villagers have the proximate ability to either protect or destroy the forests. Numerous academics and policy makers have underscored the importance of involving communities in forest management in Madagascar (e.g., Bertrand, Horning, and Montagne 2009; Horning 2012; Montagne and Ramamonjisoa 2006; Pollini and Lassoie 2011; Toillier, Lardon, and Hervé 2008). Such involvement will demand consultations that begin, rather than end, with funding and resources devoted to long-term discussions with villagers and the development of maps that recognize village resource needs, current use patterns, and community protected zones (Corson 2012). They will require mechanisms to account for internal differentiation within communities (Evers et al. 2011; Kull 2002a) and for the disconnect between conservationists' and villagers' perceptions of environmental problems (Kaufmann 2014; Keller 2008; Scales 2012). They will necessitate two-way conversation and negotiation (Richard and Dewar 2001) rather than persuasion or *sensibilisation*, and "granting equal decisional power to local groups which would include their right to oppose and even veto certain conservation measures" (Evers in Kull et al. 2010, 121–22).

As this latter idea signals, redressing Madagascar's environmental crisis requires redistributing not only resource access and control but also decision-making authority. It will entail a multipronged reconfiguration across multiple scales from Washington to Malagasy villages of the transnational power relationships through which both the current environmental crisis and the proposed solutions to it have been created. In Madagascar it will mean strengthening the national, regional, and local public institutions needed to enforce existing environmental legislation, to combat corruption, and to secure more of the wealth from natural resource exploitation that can be reinvested into the economy. In the United States it will require a lobby organization that advocates not just for species conservation but also for a more integrated and multisectoral approach to environmental conservation. At the international scale it will mean a redistribution of funds currently circulating within elite conservation circles in the conservation enterprise to the regions and villages who need it for actual environmental management.

Environmental dynamics in Madagascar cannot be addressed in isolation from the global political economy or the domestic politics of donor countries: struggles over land and resources are intertwined with both. The support of the United States for biodiversity conservation *over there* is made possible only by the American interest in saving charismatic megafauna, combined with a desire to not have conservation *over here*. Ultimately, effective conservation in Madagascar will necessitate confronting, not embracing, the larger institutionalized drivers of environmental change—the primary one of which is growing demand for and exploitation of natural resources.

How can these transformations be made? I return to my theoretical assertions that foreign aid is a terrain of contingent, temporary, and fragile articulations that occur in historically specific contexts, emerge through practice, and are subject to continual reworking. While history shapes the conditions of possibility, there are always opportunities to reconfigure alliances at particular historical conjunctures. We are at such a conjuncture now—as evidenced by a number of counterhegemonic movements around the world, including those that claim resource rights as basic human rights. This is not to naively downplay the structural power of capitalism or green neoliberalism but to underscore the ability of individuals to reshape the discursive framework through which political change takes place. If we accept that policy is made not only through formal political agreements but also through informal interactions, then we can change it through multiple avenues. Such change, however, will entail the involvement of more and different actors in transnational networks—in other words, a transformation in who travels through the corridors of power.

Appendix

List of Interviewees by Position and Organization

Agency Environmental Coordinator, U.S. Agency for International Development*

Agricultural Officer, U.S. Agency for International Development

Assistant Director, Government Relations, Audubon Society

Assistant Professor, Department of Environmental Science, Policy and Management, University of California, Berkeley

Assistant Technique, Directeur Général des Eaux et Forêts

Biodiversity and Natural Resource Specialist, U.S. Agency for International Development

Biodiversity Chair, The H. John Heinz III Center for Science, Economics and the Environment

Biodiversity Team Leader, Bureau for Economic Growth, Agriculture and Trade, U.S. Agency for International Development

Bureau Environmental Advisor, Africa Bureau, U.S. Agency for International Development

Chargé de Conservation et de Développement, Fianarantsoa, Parcs Nationaux Madagascar

Chargée de Programme, Comité Multilocal de Planification

Chargée des Opérations, Développement Rural, Environnement et Social, World Bank, Madagascar

Chef d'Unit de Production et de Diffusion des Informations Environnementales, Office National pour l'Environnement

Chef de la Circonscription, Circonscription de l'Environnement, des Eaux et Forêts, Fianarantsoa

Chef de la Circonscription, Ministère de l'Environnement, des Eaux et Forêts, Fort Dauphin

Chef de la Circonscription, Ministère de l'Environnement, des Eaux et Forêts, Toamasina

Chef de Service de Techniques Forestières, Ministère de l'Environnement, des Eaux et Forêts, Fianarantsoa

Chef de Service de Techniques Forestières, Ministère de l'Environnement, des Eaux et Forêts, Fianarantsoa

Chef du Service Planification Suivi Evaluation, Ministère de l'Environnement, des Eaux et Forêts, Direction Inter-régionale de Toamasina

Chief of Party, Jariala Project (International Resources Group)

Chief of Party, Project Director, Chemonics Madagascar, BAMEX project

Communications Director, International Conservation Caucus Foundation

Congressional Aide, Senate Appropriations Committee, Subcommittee on Foreign Operations**

Congressional Liaison, World Wildlife Fund

Conseiller Développement Rural et Environnement, Union Européenne, Délégation de la Commission Européenne à Madagascar

Conseiller Technique Forestière, Toamasina, Jariala Project (International Resources Group)

Conseiller Technique, Ambassade de France, Service de Coopération et d'Action Culturelle, Auprès du Secrétaire Général, Ministère de l'Environnement des Eaux et Forêts

Conseillère Technique Senior en Environnement et Agriculture, CARE International à Madagascar

Conservation Director, World Wide Fund for Nature Madagascar and West Indian Ocean Programme

Consultant, Equilibrium

Consultant, GTZ Programme

Consultant, Meyers Consulting

Coordinateur Adjoint, Secrétariat Multi-Bailleurs

Coordinateur d'Unité Technique Régional Toamasina, Service d'Appui à la Gestion de l'Environnement

Coordinateur Général des Projets, Ministère de l'Environnement, des Eaux et Forêts

Coordinateur National, Ecoregional Initiatives Project

Coordinateur Régional, Fianarantsoa, Ecoregional Initiatives Project

Coordinateur Régional, Fianarantsoa, Jariala Project (International Resources Group)

Coordinateur Régional, Fianarantsoa, PACT, Madagascar

Coordinateur Régional, Toamasina, Conservation International-Madagascar

Coordinateur Régional, Toamasina, Ecoregional Initiatives Project

Coordinateur Sous-Ecorégional du programme Ala-Maiky, World Wide Fund for Nature Madagascar and West Indian Ocean Programme

Coordinateur, Bailleurs de Fond, Ministère de l'Environnement, des Eaux et Forêts

Coordinateur, Commission Mines-Forêts, Jariala Project (International Resource Group)

Coordinator of the Donor Secretariat for Rural Development, World Bank, Madagascar

Coordonnateur, Cellule de Coordination du PE3, Ministère de l'Environnement, des Eaux et Forêts

Country Program Director, Wildlife Conservation Society, Madagascar

Délégué, Fondation Suisse pour le Développement et la Coopération Internationale, Madagascar

Directeur de l'Appui Technique, Ministère de l'Environnement, des Eaux et Forêts

Directeur de la Préservation de la Biodiversité, Ministère de l'Environnement, des Eaux et Forêts

Directeur de Programme à Madagascar, Durrell Wildlife Conservation Trust

Directeur de Valorisation des Ressources Forestières, Ministère de l'Environnement, des Eaux et Forêts

Directeur du Département, Evaluation Environnementale, Office National pour l'Environnement

Directeur du Développement Régional, Région Atsinanana

Directeur du Développement Régional, Région Haute Matsiatra

Directeur du Programme Terrestre, Wildlife Conservation Society, Madagascar

Directeur Exécutif, Service d'Appui à la Gestion de l'Environnement

Directeur Général Adjoint, Parcs Nationaux Madagascar

Directeur Général de l'Environnement, Ministère de l'Environnement, des Eaux et Forêts

Directeur Général des Eaux et Forêts, Ministère de l'Environnement, des Eaux et Forêts

Directeur Général, Association Nationale d'Actions Environnementales

Directeur Général, Office National pour l'Environnement

Directeur Interrégional, Ministère de l'Environnement, des Eaux et Forêts, Toamasina

Directeur Interrégional, Toamasina, Parcs Nationaux Madagascar

Directeur National, Madagascar Institut pour la Conservation des Environnements Tropicaux

Directeur Provincial d'Exécution du Projet, Toamasina, Projet de Soutien au Développement Rural

Directeur Provincial, l'Energie et des Mines

Directeur Régional du Développement Rural, Ministère de l'Agriculture

Directeur Technique, Center for Biodiversity Conservation, Conservation International-Madagascar

Directeur, CARE International à Madagascar

Directeur, Centre National de Recherches sur l'Environnement

Directeur, Institut de Civilisation, Université d'Antananarivo, Musée d'Art et d'Archéologie

Directeur, Parc National Andohahela, Fort Dauphin, Parcs Nationaux Madagascar

Director of Conservation Strategies, World Wildlife Fund

Director, Office of Agriculture, Bureau for Economic Growth, Agriculture and Trade, U.S. Agency for International Development

Director, Environmental Programs Division, Rio Tinto QIT Madagascar Minerals

Director, Office of Environment and Science Policy, U.S. Agency for International Development

Director, PACT, Madagascar

Executive Director, Bank Information Center

Expert en Forestière, Jariala Project (International Resources Group)

Fianarantsoa Sub Ecoregion Coordinator, World Wide Fund for Nature Madagascar and West Indian Ocean Programme

Field Biologist, World Wide Fund for Nature Madagascar and West Indian Ocean Programme

Financial Analyst, U.S. Agency for International Development-Madagascar

Former Administrator, U.S. Agency for International Development

Former Assistant Administrator for Economic Growth, Agriculture and Trade, U.S. Agency for International Development

Former Assistant Secretary and Counselor to the Secretary, Smithsonian Institution

Former Chief Environmental Officer, U.S. Agency for International Development

Former Congressional Aide, House Appropriations Committee**

Former Congressional Aide, House Appropriations Committee**

Former Congressional Aide, House Appropriations Committee**

Former Congressional Aide, Senate Appropriations Committee, Subcommittee on Foreign Operations**

Former Congressional Aide, Senate Appropriations Committee, Subcommittee on Foreign Operations**

Former Coordinator, NGO Outreach Project, American Farmland Trust

Former Deputy Administrator, U.S. Agency for International Development

Former Deputy Assistant Administrator, Africa Bureau, U.S. Agency for International Development

Former Directeur Général, Ministère de l'Environnement, des Eaux et Forêts

Former Director General, Office National pour l'Environnement

Former Director, Asia and Pacific Region, Biodiversity Support Program

Former Director, Office of Environment and Natural Resources, U.S. Agency for International Development

Former Director, Office of International Affairs, Natural Resources Defense Council

Former Director, Office of Natural Resource Management, U.S. Agency for International Development

Former Environment and Natural Resources Management Officer, U.S. Agency for International Development*

Former Environment and Natural Resources Management Officer, U.S. Agency for International Development

Former Environment and Natural Resources Management Officer, U.S. Agency for International Development

Former Environment and Natural Resources Management Officer, U.S. Agency for International Development

Former Environment and Natural Resources Management Officer, U.S. Agency for International Development

Former Environment and Natural Resources Management Officer, U.S. Agency for International Development-Madagascar

Former Environment and Natural Resources Management Officer, U.S. Agency for International Development-Madagascar

Former Environment Officer, U.S. Agency for International Development-Madagascar

Former Environment Officer/Consultant, U.S. Agency for International Development

Former Environment Officer, U.S. Agency for International Development

Former Legislative Affairs Liaison, Sierra Club

Former Legislative Affairs Liaison, World Wildlife Fund

Former Président, Conseil National Pour l'Environnement

Former President, Consultative Group on Biodiversity

Former Program Manager, Office of Energy, Environment & Technology, U.S. Agency for International Development

Former Senior Fellow, World Resources Institute

Head of Federal Affairs, Wildlife Conservation Society-Washington Office

Ingénieur d'Appui à la Chef de la Circonscription, Ministère de l'Environnement, des Eaux et Forêts, Haute Matsiatra

Land Resources Management Specialist, U.S. Agency for International Development-Madagascar

Land Resources Management Team Leader, U.S. Agency for International Development

Legislative and Policy Director, Senator**

Legislative Assistant, Congressional co-chair of International Conservation Caucus**

Legislative Assistant, Congressional co-chair of International Conservation Caucus**

Managing Advisor, Institut de Gemmologie de Madagascar

Managing Director, The Natural Capital Project

Marine Program Officer, World Wide Fund for Nature Madagascar and West Indian Ocean Programme

Mission Director, Madagascar, U.S. Agency for International Development

Mission Environmental Officer, U.S. Agency for International Development-Madagascar

Monitoring and Evaluation Coordinator, Conservation International-Madagascar

Natural Resource Policy Advisor, U.S. Agency for International Development

Natural Resources Management Specialist, U.S. Agency for International Development-Madagascar

Permanent Representative, Madagascar, Missouri Botanical Garden

Point Focal du Système des Aires Protégées de Madagascar, Ministère de l'Environnement, des Eaux et Forêts

President and Chief Executive Officer, International Resources Group

President, Conservation International

Priority Setting Coordinator, Miaro Program Technical Advisor, Conservation International-Madagascar

Professor, Department of Ecology and Evolution, State University of New York at Stonybrook

Program Analyst, Office of Development Planning, U.S. Agency for International Development

Program Officer, African Biodiversity Collaborative Group, World Wildlife Fund

Regional Biodiversity Synthesis Manager, Conservation International-Madagascar

Regional Coordinator East, Misonga Project, Catholic Relief Services

Regional Coordinator, ASOA, Fort Dauphin

Regional Corridor Coordinator, Fianarantsoa, Conservation International-Madagascar

Regional Representative, Fort Dauphin, U.S. Agency for International Development-Madagascar

Regional Representative, World Wide Fund for Nature Madagascar and West Indian Ocean Programme

Regional Vice President and Director, Madagascar Center for Biodiversity Conservation, Conservation International

Resident Country Director, Madagascar Operations, Millennium Challenge Corporation

Responsable de la Gestion Communautaire des Ressources Naturelles, Toamasina, Ecoregional Initiatives Project

Responsable de la Gestion Communautaire des Ressources Naturelles, Fianarantsoa, Ecoregional Initiatives Project

Secrétaire Exécutif, Fanamby

Secrétaire Général, Ministère de l'Environnement, des Eaux et Forêts

Senior Attorney, Director of the International Program, Natural Resources Defense Council

Senior Conservation Advisor, World Wide Fund for Nature Madagascar and West Indian Ocean Programme

Senior Director for International Wildlife Conservation, The National Wildlife Federation,

Senior Director for U.S. Government Affairs, Conservation International

Senior Environmental Specialist, World Bank, Madagascar

Senior Manager, Environment and Natural Resources Division, International Resources Group

Senior Natural Resources Management Advisor, U.S. Agency for International Development

Senior Official, International Conservation Caucus Foundation

Senior Policy Advisor for International Conservation, The Nature Conservancy

Senior Policy Advisor, Smithsonian Institution

Senior Scientist, National Council for Science and the Environment

Senior Vice President for Programs, American Rivers

Senior Vice President, International Government Relations, Conservation International

Social Scientist, Natural Resource Management-Biodiversity Team, U.S. Agency for International Development

Team Leader, Environment and Rural Development, U.S. Agency for International Development-Madagascar

Title II Program Specialist, Office of Health, Population and Nutrition, U.S. Agency for International Development-Madagascar

Vice Chancellor, University of Cambridge

Vice President for International Programs, Association of Public and Land-grant Universities

Visiting Scientist, Department of Biology and Environmental Science, University of Sussex

Field visits and interviews with two mayors; and members of two COBAs (community-based management authorities) and five villages in Fianarantsoa and Toamasina Provinces

*Note that these are the interviewees' titles at the time of interview and/or in some cases their former titles as offered by the interviewees. In a number of cases, interviewees participated in more than one interview. However each interviewee is only listed once.

**Names of specific senators and representatives for whom these aides work(ed) are kept confidential per agreement with interviewees.

Notes

Chapter One. Connecting Corridors

1. The park extension deadline was eventually extended to 2012; I note that while the cited legal order lists a total of 9.4 million hectares of established protected areas and priority (temporarily protected) protected areas and sustainable forest management sites, the accompanying categorical breakdown and spatial area included only 8.7 hectares.

2. Interview, August 12, 2006. The names of all interviewees and village locations are anonymous. When citing an interview, I use a general position reference for the interviewee at the time of the interview rather than an interviewee's name. Recognizing the problematic nature of the term "foreign," I use it to refer to someone not born in Madagascar. I also use "s/he" to protect the gender identity of the informant and/or the people about whom the cited person is talking. Some quotes are translated from French or Malagasy. The currency in Madagascar is called the ariary. At the time of my field research, approximately 2,000 ariary equaled one U.S. dollar. Note that I follow the English rather than the French convention for capitalization of French titles.

3. Drawing on Weiss (2005) and Lemos and Agrawal (2006), I define governance as the assemblage of informal and formal narratives, representations, logics, legal instruments, partnerships, and other mechanisms through which political actors, including states, international organizations, private companies, and nonprofits, try to influence environmental actions and outcomes.

4. I realize that "state" and "nonstate" are not homogenous and difficult to render as mutually exclusive categories. I do not go so far as James Scott (1998) in regarding the nation as being divided into state and nonstate spaces, but I have used the term "state" to refer to U.S. or Madagascar governmental or parastatal institutions, distinguishing them from nonprofit

231

and nongovernmental organizations and private mining companies, for example. These entities do have official positions and policies, and villagers and others often experience them as unified organizations. However, as Duffy (2006) notes, the NGOs working in Madagascar are far from a unified group, and in this book I focus primarily on U.S.-based, transnationally operating NGOs.

5. Earmarks are congressional requirements that an agency spend money on a specific country, organization, or issue.

6. I use the term "environmental NGOs" to refer to the broad group of NGOs that undertake environmental activities. I then distinguish between environmental advocacy groups as the groups that have traditionally used the court system to push for environmental change and that have a broad environmental focus (i.e., Sierra Club, Natural Resources Defense Council, and Environmental Defense Fund) and international conservation NGOs as those that focus on saving foreign wildlife, habitat, and biodiversity, have significant foreign programs, and receive USAID funding. However, because I focused primarily on the conservation NGOs, I do not further divide U.S. environmental NGOs, as Bramble and Porter (1992) do, into: (1) membership organizations with a broad domestic environmental focus; (2) membership organizations with an international focus; and (3) think tanks. Furthermore, I do not examine the relationship between USAID and development NGOs in depth.

7. Shifting cultivation entails a temporally and spatially cyclical clearing of land, often with the assistance of fire, followed by alternating fallow and cultivation periods (Thrupp, Hecht, and Browder 1997). Scales (2014a) defines it as "a shifting system of forest clearance for the cultivate of rainfed rice, as practiced in the eastern rainforests of Madagascar" and distinguishes it from *hatsake*, which is conducted in the west and southwest for maize, and *jinja*, which is conducted in the northwest of the island (106).

8. Authorization legislation, such as the Foreign Assistance Act, regulates what the agencies do, while the appropriations process funds programs and agencies already authorized through annual appropriations bills, such as the foreign operations bill (Epstein 2003; Streeter 2004).

9. Certain restrictions regulate how NGOs can interact with the U.S. Congress. Under section 501(c)(3) of the Internal Revenue Code (26 U.S.C. § 501(c)(3)), charitable, religious, and educational organizations that are exempt from federal income taxation (which allows them to receive private donations that are tax deductible for the contributing party) may not engage in lobbying activities that constitute a "substantial part" of their activities. The definition of "lobbying" excludes various activities, such as nonpartisan analysis, study, or research and/or advice given at the request of a governmental body. Moreover, Supreme Court rulings establish precedent that a 501(c)(3) organization can create a 501(c)(4) affiliate, which can conduct unlimited lobbying and advocacy (Maskell

1998). Thus, while many NGOs cannot directly lobby as a "substantial" part of their activities, they can influence congressional members in a variety of ways that are similar to lobbying. Moreover, a number of NGOs have also set up affiliate organizations specifically for the purposes of lobbying. Thus I use the term "lobby" in my analysis to refer to the broad range of activities they engage in.

10. Alison Jolly's archives are held in the Division of Rare and Manuscript Collections, Cornell University Library, and are open to the public. For additional information, see http://rmc.library.cornell.edu/.

Chapter Two. The History of Forest Politics in Madagascar

1. Epigraph quotes are from the following: Abel Parrot, cited in Bertrand, Alain, Jesse Ribot, and Pierre Montagne. 2004. "The Historical Origins of Deforestation and Forestry Policy in French Speaking Africa: From Superstition to Reality." In *Beyond Tropical Deforestation: From Tropical Deforestation to Forest Cover Dynamics and Forest Development*, ed. Didier Babin, 451–63, 458. Published jointly by United Nations Educational, Scientific and Cultural Organization and the Centre de Coopération International en Recherché Agronomique pour le Développement; Lavauden, Louis. 1934. "Histoire de la Législation et de l' Administration Forestière à Madagascar." *Revue des Eaux et Forêts* VIII Série 32e, Année n° 11 Novembre, 953; Sodikoff, Genese. 2005. "Forced and Forest Labor Regimes in Colonial Madagascar, 1926–1936." *Ethnohistory* 52(2): 407–35, 410; and Keller, Eva. 2008. "The Banana Plant and the Moon: Conservation and the Malagasy Ethos of Life in Masoala, Madagascar." *American Ethnologist* 35(4): 650–64, 661.

2. Article 101: The forest should not be burned and those who burn them will be put in irons for ten years.

Article 102: Charcoalers and bambooers can make their charcoal or bamboo only in denuded regions, not in the interior of the forest. Violations will be punished by a fine of three cattle and three *piastres* (currency) and, if they cannot pay, put in prison with the judgment of one *sikajy* (1/8 of a piastre) by day until paid.

Article 103: Charcoalers cannot cut big trees for making charcoal. Those who do will be punished with a fine of one cattle and one piastre per big tree, and if they cannot pay, put in prison with the judgment of one sikajy by day until paid.

Article 104: Houses cannot be constructed in the forest without authorization from the government. If people erect one for living, they will be punished by ten cattle and ten piastres, and their houses will be destroyed, and they should also pay a fee of one cattle and one piastre per tree cut. If they cannot pay, they will be put in prison with the judgment of one sikajy by day until paid.

Article 105: One cannot deforest by fire with the goal of establishing rice fields, maize, or other cultures. Only those that are already deforested and burned can be cultivated. If people use fire to create a new deforestation or extend existing cultivated areas, they will be put in irons for five days.

Article 106: The trees next to the ocean cannot be cut or damaged unnecessarily, unless ordered by the government. He who damages the forest unnecessarily will be punished with a fine of ten cattle and ten piastres, and if they cannot pay, put in prison with the judgment of one sikajy per day until paid. Translated from the *Code of 305 Articles of 1881* and from Lavauden (1934). See the United States Bureau of Manufactures (USBM 1894) for background on piastres and sikajy.

3. Following French law, a *décret* clarifies or defines how to apply a law. It is passed at the ministerial level and has to be agreed to by all ministers in the government. An *arrêté* (order) further defines the details of a law or decree.

4. The first ten included Betampona, Zahamena, Tsaratanana, Andringitra, Lokobe, Ankarafantsika, Namoraka, Tsingy du Bemaraha, Lake Tsimanampetsotsa, and Masoala. Then Andohahela was created in 1939 (Fenn 2003), and Marojejy in 1952. In 1964 Masoala was declassified as a reserve (Andriamampianina 1987).

5. The contemporary definitions of strict nature reserves, national parks, and special reserves, according to Randrianandianina et al. (2003, 1423), are as follows:

    1. A *strict nature reserve* is defined as "an area representing a particular ecosystem; its aim is to protect the fauna and flora within a boundary. To a large degree this type of reserve is an area in which nature must be left to its own devices, all human activity is forbidden, and no circulation or scientific research can be done without the permission of the appropriate authorities."

    2. A *national park* is "an area set aside to protect and preserve an exceptional natural or cultural patrimony and presenting a recreational and educational environment. Further, the intent of these parks is the protection and conservation of fauna and flora species in sites with aesthetic, geological, prehistoric, historical, archaeological, and scientific interests for the general public."

    3. A *special reserve* is "an area created mainly with the [goal] of protecting a specific ecosystem or site or a particular plant or animal species."

6. Interviews with a bilateral donor official, December 1, 2005, and senior Madagascar government officials, October 14, 2005, and November 4, 2005.

Chapter Three. Setting the Biodiversity Conservation Stage

1. Sponsors included the IUCN, UNESCO, FAO, Office de la Recherche Scientifique et Technique d'Outre-mer, International Biological Program, World Wide Fund for Nature, International Council for Bird Preservation, and the National Museum of Natural History in Paris (IUCN 1972).

2. Interview, October 10, 2005.

3. Resolutions adopted at the 1970 conference focused on the following topics: (1) the need to include nature conservation in national planning; (2) the need to plan land use on the basis of ecological principles; (3) the importance of scientific research on the conservation of natural resources; (4) education; (5) introduction of exotic species; (6) trade in endangered species; (7) forest conservation; (8) conservation of the tropics; (9) littoral zones; (10) conservation of marine tortoises; (11) threatened species; (12) national parks and other reserves; (13) association for the study and conservation of nature; and (14) the establishment of a Malagasy section of World Wide Fund for Nature (IUCN 1972, n.p.).

4. Personal communication with Alison Jolly, July 19, 2010.

5. Interview with a USAID Madagascar mission official, October 18, 2005.

6. Interview with a foreign scientist, February 27, 2010.

7. Interview, October 3, 2006.

8. Interview with senior international conservation organization representatives, October 19, 2005, and January 2, 2007.

9. Interview, September 29, 2006.

10. Interview with a former senior Madagascar government official, November 4, 2005.

11. Interview with a foreign scientist, September 24, 2006.

12. In reality the prince said this on a pivotal side trip to the Berenty reserve with Russ Mittermeier (then with WWF-U.S.) and Jean-Jacques Petter (Jolly 2015).

13. Interview with a senior international conservation NGO representative, September 29, 2006.

14. Personal communication with a senior international conservation NGO representative, December 29, 2014.

15. Interview with foreign scientists, September 24, 2006, and May 26, 2006, and a senior international conservation NGO representative, January 2, 2007.

16. These included Josef Randrianasolo, minister of livestock, water, and forests; Henri Rasolondrainibe, secretary general, Ministry of Scientific Research; Madame Berthe Rakotosamimanana, consular official, Ministry of Higher Education; Voara Randrianasolo, director of Parc Tsimbazaza; and Joelina Ratsirarson, representative, Ministry of Waters and Forests, as well as Barthélémi Vaohita.

17. Interview with a foreign scientist, September 24, 2006.

18. Personal communication with Russ Mittermeier, December 29, 2014.

19. Interviews with a foreign scientist, September 24, 2006, and a senior international conservation NGO representative, January 7, 2007.
20. In 2013 its name changed to Madagascar Fauna and Flora Group.
21. Interviews with scientists, May 26, 2006, June 16, 2006, and September 24, 2006, and a senior international conservation NGO representative, January 7, 2007.
22. Interview, May 26, 2006.
23. Interview with a foreign scientist, September 24, 2006.
24. Interview, October 10, 2005.
25. Interview with a senior international conservation NGO representative, January 2, 2007.
26. The eleven organizations included the World Bank, UNDP, UNESCO, USAID, French, German, Norwegian and Swiss Aid, World Wide Fund for Nature, and Conservation International.
27. The priorities selected in early planning meetings included (1) tools for environmental management; (2) erosion and control measures; (3) regional land use; (4) ecosystem valuation and biodiversity conservation; (5) improvement of the urban environment and its interface with the rural areas; and (6) energy–environment linkages as well as (7) environment and human health; (8) natural catastrophes; and (9) marine environment. The last three issues were later dropped in an effort to concentrate on a smaller number of priorities (Brinkerhoff and Yeager 1993).
28. Interview, October 14, 2005.
29. Interview with Alison Jolly, July 19, 2010.
30. Now called Madagascar National Parks.

Chapter Four. Tracing the Roots of Neoliberal Conservation

1. Portions of this chapter have been previously published in Corson, C. 2010. "Shifting Environmental Governance in a Neoliberal World: U.S. AID for Conservation," *Antipode* 42(3): 576–602.
2. The federal government's fiscal year begins on October 1 of the previous calendar year and ends on September 30 of the year of its title: e.g., fiscal year 2008 runs from October 1, 2007 to September 30, 2008.
3. In 2007 both the House and the Senate established new procedures for publicly disclosing earmarks and their congressional sponsors. They defined an earmark as a provision in legislation or report language that is included primarily at the request of a member and that "provides, authorizes, or recommends a specific amount of discretionary budget authority, credit authority, or other spending authority (1) to an entity, or (2) to a specific state, locality, or congressional district. However, the rules exclude funding set-asides that are selected through a statutory or administrative formula-driven or competitive award process" (Vincent and Monke 2009, 1). Thus for FY 2008 and FY 2009 the State, Foreign

Operations, and Related Programs appropriations subcommittee reported no earmarks (Vincent and Monke 2009).

4. Interview with a senior environmental NGO representative, June 30, 2006.

5. A number of projects in soil conservation, fuelwood production and conservation, and other related issues, all of which one could call environmental projects, preceded this legislation. However, this period marks the beginning of the agency's common use of the descriptor "environmental."

6. Other parties to the suit were the National Audubon Society and the Sierra Club.

7. Interview, January 8, 2007.

8. Interview, June 30, 2006.

9. Interview, June 29, 2006.

10. Interview, June 21, 2006.

11. Interview, January 10, 2007.

12. Interview, July 31, 2008.

13. Interview with a former NGO representative, January 8, 2007.

14. Interviews with a senior environmental advocacy NGO representative, June 30, 2006; former and current USAID officials, June 21, 2006, and January 2, 2007; and a congressional aide, January 10, 2007.

15. Interview, August 3, 2008.

16. Interview with a senior environmental NGO representative, June 30, 2006.

17. Interview, January 10, 2007.

18. According to USAID, the initialism "PVO" includes both U.S.-based and indigenous private voluntary organizations. It also includes cooperative development organizations (CDOs). "NGO" refers to nongovernmental organizations, including PVOs and CDOs unless otherwise specified (USAID 2002b). The U.S. Government Accounting Office (GAO) provides a more detailed definition: "PVOs are tax-exempt, nonprofit organizations that receive voluntary contributions of money, staff time, or in-kind support from the general public and are engaged in voluntary, charitable, or development assistance activities. PVOs and NGOs can be U.S. based, international, or locally-based in the host country" (GAO 2002, 4).

19. In practice this meant "U.S. NGOs." Several bureaucratic impediments mean that it is difficult for non-U.S. NGOs to qualify for USAID funding. First, USAID can only contract to organizations on its accepted roster of potential grantees and contractors. Second, USAID's registration and reporting requirements are onerous. For example, all USAID grants and contracts must be audited by one of the major accounting firms (Smillie 1999).

20. Interview, January 10, 2007.

21. Interview, June 22, 2006.

22. Interview, June 23, 2006.
23. Interview, August 3, 2005.
24. Interview with a USAID official, August 21, 2005.
25. Interview, June 30, 2006.
26. Interview, January 10, 2007.
27. Interviews with a senior environmental NGO representative, January 2, 2007, and a congressional aide, January 10, 2007.
28. Interview, August 5, 2005.
29. Interview, June 30, 2006.
30. Interview, January 2, 2007.
31. Interview, August 8, 2005.
32. Interview, August 5, 2005.
33. Interview, January 2, 2007.
34. This report included PVOs, consulting firms, and universities as NGOs.
35. Interview with a former USAID official, August 3, 2005.
36. Interview with a USAID official, August 5, 2005.
37. Interviews with former USAID staff, August 8, 2005; current USAID officials June 26, 2006, and January 4, 2007; and a congressional aide, August 21, 2008.
38. Strategic objectives are overarching goals, usually sector specific, under which projects are categorized.
39. Interviews with USAID officials, August 19, 2005, and June 23, 2006, and a USAID contractor, August 24, 2005.
40. Interview, June 30, 2006.
41. The other three subindicators related to improved sanitation and water and child mortality.
42. Interview, June 30, 2006.
43. Interview with a USAID official, June 30, 2006.
44. Interview with a former USAID official, June 30, 2006.
45. Interview, June 23, 2006.
46. Interview, January 2, 2007.
47. Interview with a USAID contractor, August 24, 2005.
48. Interview with a former senior USAID official, January 2, 2007.
49. Interview with a former USAID Madagascar official, June 15, 2006.
50. Interviews with a USAID Madagascar mission official, December 13, 2005; a senior international conservation NGO representative, December 3, 2005; and a senior USAID Madagascar contractor, December 3, 2005.
51. Interview, August 4, 2005.
52. Interview, June 22, 2006.
53. Interview with a congressional aide, August 9, 2005.
54. Interviews with USAID officials, August 19, 2005, June 23, 2006, and January 4, 2007.
55. Interview, January 10, 2007.

56. Interview, June 23, 2006.
57. Interview, June 23, 2006.
58. Interview, June 16, 2006.
59. Interviews with USAID officials and contractors, October 18, 2005, June 22, 2006, and January 4, 2007.
60. Interview, August 5, 2005.
61. Interview with a USAID official, January 4, 2007.
62. While I focus on the biodiversity earmark, additional congressional mandates had similar effects on the environmental portfolio. For example, beginning in 2008 the appropriations bill included an earmark for US$300 million for safe drinking water and sanitation supply projects to implement the Senator Paul Simon Water for the Poor Act of 2005 (Public Law 109–121) and an earmark for US$195 million for clean energy and other climate change programs (U.S. Congress 2007).
63. Interviews with conservation NGO congressional liaisons, August 19, 2005, June 16, 2006, June 22, 2006, January 5, 2007, January 9, 2007, and a congressional aide, January 8, 2007.
64. Of note, congressional directives plateaued around US$200 million: while the four conservation NGOs pushed for US$250 million in FY 2009 and US$350 million for FY 2011, for those fiscal years congressional directives set aside US$200 million and US$215.2 million respectively (U.S. Congress 2008, 2010).
65. A congressional caucus is a group of congressional members who convene to pursue common legislative objectives.
66. Interviews with a conservation NGO congressional liaison, June 16, 2006, and NGO senior staff, June 30, 2006.
67. Interview, August 3, 2005.
68. Interview, January 5, 2007.
69. Interview, June 16, 2006.
70. Interviews with a conservation NGO congressional liaison, June 22, 2006, and a former congressional aide, June 29, 2006.
71. Interview, January 5, 2007.
72. After a number of exposés published in *Mother Jones* magazine in 2013 and 2014, WWF-US and CI left the ICCF (Hiar 2014).
73. Interview, January 9, 2007.

Chapter Five. A Model for Greening Development

1. Interview, July 7, 2006.
2. Interviews with Madagascar government officials, October 8, 14, and 28, 2005; senior international conservation NGO representatives, November 3 and 25, 2005; a bilateral donor official, December 1, 2005; a senior USAID Madagascar contractor, October 24, 2005; and USAID Madagascar mission official, November 18, 2005.

3. A French term that might be translated into English as "persuasive education," "outreach," or "awareness-raising."

4. In 1996 the National Assembly impeached Zafy Albert on grounds of corruption and exceeding constitutional power. In elections that followed, Didier Ratsiraka won a second term and stayed until his defeat in December 2001 by Marc Ravalomanana. "La crise," or the "2002 crisis," that followed this election entailed several months of disputes over the victory in which Ratsiraka refused to step down, multilateral and bilateral donors left the country, conflicts broke out across the country, and over one hundred thousand supporters of Ravalomanana took to the streets of Antananarivo. On February 22, 2002, Ravalomanana declared himself president and eventually Ratsiraka fled for France on June 15, 2002 (Marcus 2004).

5. Interview, January 7, 2007.

6. Interview, November 24, 2005.

7. These programs encompassed the following: sustainable soil and water management; multiple-use forest ecosystem management; national parks and ecotourism; marine and coastal environment; local natural resource management and land tenure security; regional programming and spatial analysis; regional fund for environmental management; upgrading of the legal framework and formulation of environmental policies; assisting sector ministries in implementing policies and making environmental impact assessment operational; research; education and training; geographic instruments; environmental information system; and communication, monitoring, evaluation, program coordination, and management (World Bank 2007). Core donors included the World Bank; United States, French, German, and European aid agencies; WWF; CI; International Fund for Agricultural Development; Global Environment Facility (GEF); and UNDP (Froger and Andriamahefazafy 2003).

8. The seven AGEX were ANAE, ANGAP, ONE, SAGE, Centre de Formation aux Sciences de l'Information Géographique et de l'Environnement (Environment and Geographic Information Training Center), Institut Géographique de Madagascar (Madagascar Geographic Institute), and the Forest Service.

9. Interviews, November 4 and 24, 2005.

10. Interviews with a senior Madagascar government official, November 4, 2005, and a USAID Madagascar official, November 24, 2005.

11. Interviews with a former senior Madagascar government official, October 10, 2005; a Madagascar government official, October 8, 2005; a Malagasy NGO representative, November 2, 2005; and a senior Madagascar government official, November 4, 2005.

12. Interview with a Madagascar government official, November 24, 2005.

13. Key donors to EP3 included the World Bank, the UNDP, CI, WWF, WCS, the European Union, KfW, and Deutsche Gesellschaft für Technische

Zusammenarbeit, the Swiss Agency for Development and Cooperation, the French Department for International Cooperation, and the Japan International Cooperation Agency. The World Bank and USAID each pledged US$35 million, KfW gave US$20 million, the Global Environment Facility gave US$8 million, the UNDP funded US$4 million, and other bilateral donors provided about US$6.4 million (Lindemann 2004, 9).

14. Interview with a Madagascar government official, October 20, 2005.
15. Interviews with a USAID Madagascar mission official and an official from another bilateral donor.
16. Interview with a foreign scientist, September 15, 2006.
17. Interview, November 4, 2005.
18. Interview, October 28, 2005.
19. A decree of February 2007 combined the directorates of Environment, Water, and Forests in one ministry and created 22 regional directorates of the ministry in addition to the 107 district representatives (OSF 2007).
20. Interview with a regional Madagascar government official, August 7, 2006.
21. Interviews with a senior development NGO representative, November 7, 2005, a development NGO representative, December 1, 2005, and a senior USAID Madagascar contractor, October 17, 2005.
22. Interview with a senior USAID Madagascar contractor, October 17, 2005.
23. Interview with a development NGO representative, December 1, 2005.
24. Interview, December 8, 2005.
25. Interview with Madagascar government official, November 11, 2005.
26. The large-scale mining sector includes industrial minerals and ores, such as gold, graphite, chromite, mica, coal, iron, ilmenite, nickel, bauxite, titanium oxide, beryl, quartz limestone, cement, and alumina.
27. These include diamonds, rubies, emeralds, aquamarine, tourmaline, and amethyst as well as decorative stones such as labradorite, rock crystal, rhodonite, marble, cordierite, celestite, vitreous beryl, quartz, opaque tourmaline, corundum, ammonite, aragonite; silicified wood; and gold.
28. Interviews with senior Madagascar government officials, December 14, 2005, and September 14, 2006; and a senior USAID Madagascar contractor, July 26, 2006.
29. Interviews with a senior USAID Madagascar contractor, October 21, 2005; a regional Madagascar government official, October 21, 2005; and a regional Madagascar government official, November 14, 2005.
30. Interviews with a USAID Madagascar contractor, September 19, 2006, and a regional Madagascar government official, August 14, 2006.
31. Interviews with a regional Madagascar government official, November 11, 2005; an international conservation NGO representative, September 13, 2006; and a regional Madagascar government official, November 14, 2005.

32. Interview, November 11, 2005.

33. The MAP laid out "direction and priorities for the nation from 2007 to 2012 [to] ignite rapid growth, lead to the reduction of poverty, and ensure that the country develops in response to the challenges of globalization and in accordance with the national vision—Madagascar Naturally! and the United Nations Millennium Development Goals." Its priorities included "responsible governance"; "connected infrastructure"; "educational transformation"; "rural development and a green revolution"; "health, family planning and the fight against HIV / AIDS"; "high growth economy"; "cherish the environment"; and "national solidarity" (Repoblikan'i Madagasikara 2006a, 3).

Chapter Six. Creating the Transnational Conservation Enterprise

1. For example, interviews with an international conservation NGO representative, December 1, 2005, and a senior Madagascar government official, October 28, 2005.

2. Interview, October 24, 2005.

3. The Agricultural Trade Development and Assistance Act of 1954 (P.L. 480), commonly known as Food for Peace, established the basis for U.S. foreign food assistance. There are four relevant sections of the act: Title I relates to government-to-government sales under long-term credit arrangements of agricultural products to developing countries; Title II covers the donation of agricultural commodities for humanitarian purposes; Title III provides for government-to-government grants for economic development; Title V provides agricultural technical assistance. USAID administers titles II and III (USAID 2006; USDA 2008).

4. Interview with Samuel S. Rea as part of the Foreign Affairs Oral History Collection of the Association for Diplomatic Studies and Training, interviewed by W. Haven North, April 24, 1998.

5. Ibid.

6. WCS and CI did not start working in Madagascar until the 1990s.

7. Budget presentations and Country Development Strategy Statements were usually developed two years prior to their proposed fiscal implementation year.

8. Interview, January 2, 2007.

9. Interview, September 15, 2006.

10. In theory this meant U.S.- and Malagasy-based contractor and grantee organizations, but in practice it increased USAID funding to U.S.-based organizations.

11. The project reflected a move in the 1980s and 1990s to provide "non-project assistance," such as balance-of-payments support, in exchange for policy reforms (Bratton 1990).

12. ARD was acquired by Tetra Tech in 2007.

13. Several notes on USAID's EP1 programs: (1) MBG managed Masoala until CARE took over its management under the SAVEM project. World Wildlife Fund managed Andohahela Reserve. Volunteers in Technical Assistance managed Andasibe-Mantadia, with input from Clark University, the Malagasy Lutheran Church, Sampan' Asa Momba ny Fampandrosoana, and the Tropical Forestry Management Trust. The State University of New York at Stony Brook took over management of Ranomafana National Park from Duke University under the SAVEM project. Cornell University and the Malagasy NGO Tery Saina managed agriculture activities under the project. CI managed the Zahamena Strict National Reserve. WWF-US, CARE, and Veterinarians without Borders managed Amber Mountain / An karana / Analemera. (2) Between EP1 and the second phase of the Madagascar EP2, a bridge project entitled Managing Innovative Transition in Agreement provided further assistance to ANGAP (Freudenberger 2010). (3) The USAID Washington office also directly funded projects in Madagascar: First, through a cooperative grant agreement with U.S.-based NGOs (WWF, Experiment in International Living, and CARE), it gave US$230,000 to a consortium of NGOs entitled Le Conseil Malgache des Organisations Non-Gouvernementales pour le Développement et l'Environnement (COMODE) (Malagasy Council of NGOs for Environment and Development) for program management, consultations, and technical assistance (USAID 1994a). Officially created in December 1990, COMODE comprised about thirty NGO members, among them the major religiously affiliated development NGOs active in Madagascar (Greve 1991a; USAID 1994a). And, second, through BSP, which funded WWF, WRI, and TNC, small grants for biodiversity and the USAID / National Science Foundation Collaborative Program in Biological Diversity, a U.S. Congress–mandated collaborative program in biological diversity research, and project NOAH, which funded projects that furthered preservation of genetic material (USAID 1994a).

14. Letter from Carlos Gallegos, Chief, Office of Natural Resources, USAID-Madagascar to Russ Mittermeier, President, CI, May 25, 1994.

15. Letter to George Carner, Director, USAID-Madagascar, June 23, 1994, signed by the following: David F. Anderson, Director, San Francisco Zoological Gardens; John Behler, Curator, Department of Herpetology, New York Zoological Gardens; John Hartley, Assistant Director, Jersey Wildlife Preservation Trust; Rick Hudson, Asst. Reptile Curator, Fort Worth Zoological Park; Alison Jolly, Visiting Lecturer, Princeton University; Paul V. Loiselle, Curator of Freshwater Fishes, Wildlife Conservation Society; Porter S. Lowry III, Head of Africa Program, Missouri Botanical Garden; Russell A. Mittermeier, President, Conservation International; Ronald A. Nussbaum, Curator of Amphibians and Reptiles, Museum of Zoology, University of Michigan; George B. Rabb, Chairman, IUCN Species Survival Commission, Chicago Zoological

Park, Brookfield Zoo; Christopher Raxworthy, Research Investigator, Division of Herpetology, University of Michigan; Simon Stuart, Head, Species Survival Program, IUCN; Jorgen B. Thompson, Director, TRAFFIC International; and Sukie Zeeve, Project Coordinator, MFG.

16. Interview with a USAID Madagascar mission official, November 18, 2005.

17. Interviews with former and current USAID Madagascar officials, November 29, 2005, and May 26, 2006.

18. Interview, November 29, 2005.

19. Note that LDI was followed by a Programme de Transition Eco-Régional from 1998 until 2004, which "allowed LDI to continue operating until the new ERI program was signed" (Freudenberger 2010, 16).

20. During the SAPM process the boundaries of the first two corridors were expanded northward and southward, but at the time of the research they were referred to as the Ankeniheny–Zahamena and Fandriana–Vondrozo corridors.

21. *Jariala* is a Malagasy word that means "forest management."

22. *Miaro* means "protection" in Malagasy.

23. Interviews with a senior USAID Madagascar contractor, December 7, 2005; a senior development NGO representative, November 7, 2005; and a USAID Madagascar mission official, September 19, 2006.

24. Interviews with a senior USAID Madagascar contractor, December 7, 2005, and a senior development NGO representative, November 7, 2005.

25. Interviews with a USAID Madagascar contractor, September 19, 2006, and a regional MinEnvEF official, August 14, 2006.

26. There were also two environment-related programs funded by USAID-Washington, with limited oversight by the USAID-Madagascar mission. The first was a grant of around US$200,000 to CI under the USAID-Washington-based Global Conservation Program's cooperative agreement, with a subgrant to Fanamby (a Malagasy NGO) and Durrell Wildlife Trust to develop a management system for the new Menabe protected area (ARD 2008; CI 2006; Freudenberger 2010). The second was a Global Development Alliance program among the Regional Authority of Anosy, USAID, and QMM to undertake a program known as the Anosy Development Alliance. The program aimed to link development and environment by supporting (1) development of the economic sector; (2) natural resource management and conservation; (3) health promotion and improvement; and (4) development of regional education opportunities (U.S. Embassy-Madagascar 2006).

27. Interviews with a USAID Madagascar mission official, September 25, 2006, and a regional development NGO representative, December 8, 2005.

28. Interview, December 1, 2005.

29. Interview, November 18, 2005.

30. Interview, October 24, 2005.
31. Interviews with USAID Madagascar mission officials, November 29, 2005, and September 23, 2006.
32. Interviews with a USAID official, October 18, 2005; a USAID contractor, July 7, 2006; and an NGO representative, June 22, 2006.
33. Interview, January 2, 2007.
34. Interview, September 23, 2006.
35. Ibid.
36. Interviews with USAID Madagascar mission officials, October, 18, 2005, and September 28, 2006.
37. Interviews with USAID Madagascar mission officials, September 11, 2006, and September 28, 2006.
38. Interview, August 1, 2006.
39. Interview with a USAID Madagascar mission official, September 28, 2006.
40. Interviews with USAID Madagascar mission officials, December 10, 2005, July 3, 2006, and September 7, 2006.
41. Interview with a USAID Madagascar mission official, August 1, 2006.
42. Interview, November 18, 2005.
43. Interview with a bilateral donor official, October 27, 2005.
44. Interview with a senior international conservation NGO representative, November 3, 2005.
45. Interviews with a senior international conservation NGO representative, January 7, 2007, and a USAID Madagascar official, November 24, 2005.
46. Interviews with USAID Madagascar mission officials, November 18, 2005, and December 3, 2005.
47. Interview with a senior international conservation NGO representative, December 3, 2005.
48. Ibid.
49. Interview with a USAID Madagascar mission official, August 1, 2006.
50. Interviews with former and current USAID Madagascar officials, December 10, 2005, and July 6, 2006.
51. Interview with a USAID Madagascar mission official, October 18, 2005.
52. Interview with a USAID Madagascar mission official, December 13, 2005.

Chapter Seven. Accountability and Authority in Conservation Politics

1. Every ten years the World Parks Congress convenes to set the agenda for international protected areas policy.
2. The most recent version of the IUCN's system of categories for protected areas includes seven categories of protected areas: Category Ia: Strict

nature reserve / wilderness: for science or wilderness protection; Category Ib: Wilderness area: for wilderness protection; Category II: National park: for ecosystem protection and recreation; Category III: Natural monument: for conservation of specific natural features; Category IV: Habitat / species management area: for conservation through management intervention; Category V: Protected landscape / seascape: for landscape / seascape conservation or recreation; Category VI: Managed resource protected area: for the sustainable use of natural resources (Dudley and Phillips 2006).

3. Portions of this chapter have been previously published in Corson, C. 2011. "Territorialization, Enclosure and Neoliberalism: Nonstate Influence in Struggles over Madagascar's Forests," *Journal of Peasant Studies* 38:4, 703–26 (reprinted in N. Peluso and C. Lund, eds. *New Frontiers of Land Control* (Routledge); and Corson, C. 2012. "From Rhetoric to Practice: How High-Profile Politics Impeded Community Consultation in Madagascar's New Protected Areas," *Society and Natural Resources* 25:4, 336–35, both of which are available at www.tandfonline.com; and in Corson, C. 2014. "Conservation Politics in Madagascar: The Expansion of Protected Areas," in *Conservation and Environmental Management in Madagascar*, 193–215, ed. I. Scales (Earthscan).

4. Interviews with international conservation NGO representatives, October 24, 2005, and September 29, 2006; a bilateral donor representative, October 27, 2005; and a former senior Madagascar government official, September 15, 2006.

5. Interview, September 15, 2006.

6. Interview with a senior international conservation NGO representative, September 29, 2006.

7. Interview, October 24, 2005.

8. Interview, September 14, 2006.

9. Interviews with a senior international conservation NGO representative, November 3, 2005, and an international conservation NGO representative, October 24, 2005.

10. Interviews with senior international conservation NGO representatives, November 3, 2005, and December 1, 2005.

11. Interview, January 2, 2007.

12. The Durban Vision Group oversaw five technical groups: (1) management and categorization, which was responsible for setting up a management system coherent with the IUCN guidelines; (2) biodiversity prioritization, which continued the biodiversity prioritization work; (3) communication, which coordinated communication with regional and central authorities and the general public; (4) legal framework, which developed legislation relative to the program; and (5) funding.

13. Interview, October 28, 2005.

14. Interview, January 2, 2007.

15. Interviews with a senior conservation NGO representative, October 19, 2005; a bilateral donor official, November 3, 2005; and a foreign scientist, March 17, 2006.
16. Interview with an international conservation NGO representative, October 19, 2005.
17. Interviews with an international conservation NGO representative, November 7, 2005, and a foreign scientist, March 17, 2006.
18. Interviews with a mining company agent, December 13, 2005, and an international conservation NGO representative, October 19, 2005.
19. Interviews with contractors, July 26, 2006, and September 11, 2006, and a regional Madagascar government official, August 22, 2006.
20. Interview with a regional international conservation NGO representative, August 18, 2006.
21. Interview, September 13, 2006.
22. Interview, September 15, 2006.
23. This can be translated as "tending the forest."
24. Interviews with a USAID Madagascar contractor, September 11, 2006, and a regional international conservation NGO representative, December 9, 2005.
25. Interview, September 20, 2006. Note that in the Mikea forest Tucker similarly found that maps presented the area as being uninhabited, ignoring the population that lives in the forest (Kull et al. 2010).
26. Interviews with senior international conservation NGO representatives, December 1, 2005, and September 12, 2006.
27. Interview, September 19, 2006.
28. Interview, November 11, 2005.
29. Interview, November 24, 2005.
30. Interviews with an international conservation NGO representative, November 24, 2005, and a USAID Madagascar mission official, December 13, 2005.
31. Interview with an international conservation NGO representative, November 7, 2005.
32. Interview, November 28, 2005.
33. Interviews with an international conservation NGO representative, November 28, 2005; a USAID Madagascar mission official, November 24, 2005; a senior USAID Madagascar contractor, October 24, 2005; and a senior international conservation NGO representative, December 3, 2005.
34. Note that I had intended to conduct extensive focus group interviews, but the villagers with whom I spoke had so little knowledge that I decided to concentrate on the challenges faced by regional staff tasked with implementing the program.
35. Interview, November 28, 2005.
36. Interview, September 6, 2006.

37. Interviews with villagers, August 12, 2006, and regional contractors, December 7, 2005, and August 7, 2006.
38. Village focus groups, August 12, 2006.
39. Interview, November 14, 2005.
40. Interview, December 8, 2005.
41. Interview, August 1, 2006.
42. Interview, November 11, 2005.
43. Interview, December 7, 2005.
44. Interview, November 11, 2005.
45. Interview with regional contractors, November 10 and 11, 2005, and August 22, 2006.
46. Interviews with a regional government official, August 14, 2006, and a regional contractor, August 7, 2006.
47. Pollini (2011) reports that a September–October 2006 supervision mission for World Bank funding of the third phase of the NEAP pointed out that insufficient funding was available for development activities around protected areas, which put the World Bank at risk of being criticized by human rights lobbies for violating its safeguards policy. He writes, "This problem was intensely debated during the EP3 joint committee of June 12, 2006," and eventually government policy recommended funding types of development projects that had already been implemented, such as community conservation, capacity building, reforestation, and microcredit (many of which, he suggests, were not benefiting communities) rather than requiring direct compensation to communities for the costs they incurred (81).
48. This would meet the newly negotiated CBD target to protect, by 2020, at least 17 percent of terrestrial and inland water and 10 percent of coastal and marine areas.

Chapter Eight. Transforming Relations of Power in Conservation Governance

1. Epigraphs are from the following: (1) http://blog.conservation.org/2014 /03/is-there-new-hope-for-madagascar/ (accessed April 19, 2014); and (2) Freudenberger, Karen. 2010. *Paradise Lost? Lessons from Twenty-Five Years of USAID Environment Programs in Madagascar.* USAID Publication (Washington, D.C.: International Resources Group), 85.
2. Interview, December 7, 2005.
3. Interviews with a Madagascar government official, November 29, 2005; a senior NGO representative, September 15, 2006; a USAID Madagascar mission official, September 23, 2006; and a regional Madagascar government official, September 25, 2006.

4. Interviews with former senior Madagascar government officials, July 31, 2006, and September 15, 2006; and a Malagasy NGO representative, November 2, 2005.

5. Interviews with a regional Madagascar government official, November 11, 2005; and Madagascar government officials, November 24, 2005, and August 1, 2006.

References

Adams, William, and David Hulme. 2001. "Changing Narratives, Policies and Practices in African Conservation." In *African Wildlife and Livelihoods*, ed. David Hulme and Marshall Murphree, 9–23. Portsmouth, N.H.: Heinemann.

Adams, William M. 2008. *Green Development: Environment and Sustainability in the Third World*. 3rd ed. London: Routledge.

Adams, William M., and Jon Hutton. 2007. "People, Parks and Poverty: Political Ecology and Biodiversity Conservation." *Conservation and Society* 5(2): 147–83.

African Convention on the Conservation of Nature and Natural Resources. 1968. Deposited with the Organization of African Unity. 16 June 1969.

Alcorn, Janice. 2005. "Dances Around the Fire: Conservation Organizations and Community-Based Natural Resource Management." In *Communities and Conservation: Histories and Politics of Community-Based Natural Resource Management*, ed. J. Peter Brosius, Anna Lowenhaupt Tsing, and Charles Zerner, 37–68. New York: AltaMira Press.

Alpert, Peter. 1996. "Integrated Conservation and Development Projects." *BioScience* 46(11): 845–55.

Anderson, Jon, Asif Shaikh, Chris Barrett, Peter Veit, Jesse Ribot, Bob Winterbottom, Mike McGahuey, and Roy Hagen. 2002. *Nature, Wealth, and Power: Emerging Best Practice for Revitalizing Rural Africa*. Washington, D.C.: USAID, in collaboration with CIFOR, Winrock, WRI, and IRG under the Environmental Policy and Institutional Strengthening Indefinite Quantity (EPIQ), Contract No. PCE-I-00-96-00002-00.

Anderson, Tom. 2013. "Solving Madagascar: Science, Illustrations, and the Normalizing of Fauna of Nineteenth Century Madagascar." In *Contest for Land in Madagascar: Environment, Ancestors, and Development*, ed. Sandra J. T. M. Evers, Gywn Campbell, and Michael Lambek, 97–118. Leiden: Brill.

Andriamahefazafy, Fano, and Philippe Méral. 2004. "La Mise en Œuvre des Plans Nationaux d'Action Environnementale: Un Renouveau des Pratiques des Bailleurs de Fonds?" *Mondes en Développement* 127(3): 29–44.

Andriamampianina, Joseph. 1987. *Statut des Parcs et Réserves de Madagascar.* Antananarivo, Madagascar. IUCN Species Survival Commission.

Andriamialisoa, Fanja, and Olivier Langrand. 2003. "History of Scientific Exploration." In *The Natural History of Madagascar,* ed. Steven Goodman and Jonathan Benstead, 1–15. Chicago: University of Chicago Press.

Andrianandrasana, Onimandimbisoa, Thomas K. Erdmann, Mark S. Freudenberger, Vololoniaina Raharinomenjanahary, Mamy Rakotondrazaka, and Jean-Solo Ratsisompatrarivo. 2008. *Le Transfert de Gestion des Ressources Naturelles Pour la Sauvegarde des Corridors Forestiers: La Vision du Programme "Eco-Regional Initiatives," à Travers Son Expérience dans les Ecorégions de Fianarantsoa et Toamasina.* Report prepared for the 2008 USAID-Madagascar Stock Taking Exercise.

ANGAP. 2003. *Madagascar Protected Area System Management Plan, Revised.* Association Nationale pour la Gestion des Aires Protégées. Antananarivo, Madagascar. ANGAP.

ARD. 1997. *Knowledge and Effective Policies for Environmental Management Final Report.* Burlington, Vt. Submitted by Associates in Rural Development, Inc., Contract #687-0113-C-00-4053-00.

———. 2008. *Evaluation of the Global Conservation Program Final Evaluation Report.* Prepared for the United States Agency for International Development, USAID Contract Number EPP-I-00-060008, Task Order No. 1.

Arsel, Murat, and Bram Büscher. 2012. "Nature Inc.: Changes and Continuities in Neoliberal Conservation and Market-Based Environmental Policy." *Development and Change* 43(1): 53–78.

Attwell, C. A. M., and F. P. D. Cotterill. 2000. "Postmodernism and African Conservation Science." *Biodiversity Conservation* 9(5): 559–77.

Atwood, J. Brian. 1993. *Statement of Principles on Participatory Development.* Washington, D.C.: USAID.

Auer, Matthew R. 1998. "Agency Reform as Decision Process: The Reengineering of the Agency for International Development." *Policy Sciences* 31(2): 81–105.

Aufderheide, Pat, and Bruce Rich. 1988. "Environmental Reform and the Multilateral Banks." *World Policy Journal* 5(2): 301–21.

Bailey, Jodi. 2006. "The Limits of Largess: International Environmental NGOs, Philanthropy and Conservation." PhD diss., Department of Geography, University of California, Berkeley.

Bakker, Karen. 2010. "The Limits of 'Neoliberal Natures': Debating Green Neoliberalism." *Progress in Human Geography* 34(6): 715–35.

Barrett, Christopher B. 1994. "Understanding Uneven Agricultural Liberalisation in Madagascar." *Journal of Modern African Studies* 32(3): 449–476.

BATS. 2008. Madagascar Environmental Threats and Opportunities Assessment, 2008 Update. In *EPIQ IQC Contract No. EPP-I-00-03-00014-00, Task Order 02:* Biodiversity Analysis and Technical Support for USAID / Africa (Chemonics International Inc., IUCN, WWF, and International Program Consortium). In coordination with program partners: the U.S. Forest Service / International Programs and the Africa Biodiversity Collaborative Group.

Bebbington, Anthony, and Uma Kothari. 2006. "Transnational Development Networks." *Environment and Planning A* 38(5): 849–66.

Bebbington, Anthony, David Lewis, Simon Batterbury, Elizabeth Olson, and M. Shameem Siddiqi. 2007. "Of Texts and Practices: Empowerment and Organisational Cultures in World Bank–Funded Rural Development Programmes." *Journal of Development Studies* 43(4): 597–621.

Becker, Dennis, Martin J. Myers, and Oliver Pierson. 2005. *Evaluation of the Current Forestry Law Enforcement Program and Development of a New Forestry Control Program Targeting Illegal Logging, Draft Report #2.* Antananarivo, Madagascar. USDA–Forest Service Technical Assistance Mission, in Support to the Direction Générale des Eaux et Forêts, USAID Madagascar, and the Jariala Program. Mission Dates: July 6–22, 2005.

Bekhechi, Mohammad, and Jean Roger Mercier. 2002. *The Legal and Regulatory Framework for Environmental Impact Assessments: A Study of Selected Countries in Sub-Saharan Africa.* Rudolf V. Van Puymbroeck. Law, Justice and Development Series. Washington, D.C. World Bank.

Belvaux, Eric. 2006. "Décentralisation et Gestion des Ressources Naturelles dans un Pays en Développement: l'Exemple de Madagascar." GECOREV Symposium on Co-management of Natural Resources and the Environment: For Dialogue Between Researchers, Civil Society, and Decision Makers, University of Versailles Saint Quentin-en Yvelines.

Berg, Robert, and David Gordon. 1989. *Cooperation for International Development: The United States and the Third World in the 1990s.* Boulder: Lynne Reinner.

Berríos, Rubén. 2000. *Contracting for Development: The Role of For-Profit Contractors in U.S. Foreign Development Assistance.* Westport, Conn.: Praeger.

Bertrand, Alain. 2004. "The Spread of the Merina People in Madagascar and Natural Forest and Eucalyptus Stand Dynamics." In *Beyond Tropical Deforestation: From Tropical Deforestation to Forest Cover Dynamics and Forest Development,* ed. Didier Babin, 151–56. Paris: UNESCO and the CIRAD.

Bertrand, Alain, and Désiré Randrianaivo. 2003. "Tavy et Déforestation." In *Déforestation et Systèmes Agraires à Madagascar: Les Dynamiques des Tavy sur la Côte Orientale*, ed. Sigrid Aubert, Serge Razafiarison, and Alain Bertrand, 9–30. Montpellier: CIRAD-CITE-FOFIFA.

Bertrand, Alain, and Rivo Ratsimbarison. 2004. "Deforestation and Fires: The Example of Madagascar." In *Beyond Tropical Deforestation: From Tropical Deforestation to Forest Cover Dynamics and Forest Development*, ed. Didier Babin, 79–87. Paris: UNESCO and the CIRAD.

Bertrand, Alain, Jesse Ribot, and Pierre Montagne. 2004. "The Historical Origins of Deforestation and Forestry Policy in French-Speaking Africa: From Superstition to Reality." In *Beyond Tropical Deforestation: From Tropical Deforestation to Forest Cover Dynamics and Forest Development*, ed. Didier Babin, 451–63. Published jointly by UNESCO and the CIRAD.

Bertrand, Alain, Nadia Rabesahala Horning, and Pierre Montagne. 2009. "Gestion Communautaire ou Préservation des Renouvelables: Histoire Inachevée d'une Evolution Majeure de la Politique Environnementale à Madagascar." *VertigO* 9(3): 1–18.

Bertrand, Alain, Sigrid Aubert, Pierre Montagne, Alexio Clovis Lohanivo, and Manitra Harison Razafintsalama. 2014. "Madagascar, Politique Forestière: Bilan 1990–2013 et Propositions." *Madagascar Conservation and Development* 9(1): 20–30.

Bhatnagar, Bhuvan, and Aubrey C. Williams. 1992. *Participatory Development and the World Bank: Potential Directions for Change*. World Bank Discussion Papers. Washington, D.C.: World Bank.

Billo, Emily, and Alison Mountz. 2015. "For Institutional Ethnography: Geographical Approaches to Institutions and the Everyday." *Progress in Human Geography* doi: 10.1177/0309132515572269.

Birchard, Bill. 2005. *Nature's Keepers: The Remarkable Story of How the Nature Conservancy Became the Largest Environmental Organization in the World*. San Francisco: Jossey-Bass.

Blake, Robert O., Barbara J. Lausche, S. Jacob Scherr, Thomas B. Stoel Jr., and Gregory A. Thomas. 1980. *Aiding the Environment: A Study of the Environmental Policies, Procedures, and Performance of the U.S. Agency for International Development*. Washington, D.C.: NRDC.

Blanc-Pamard, Chantal. 2009. "The Mikea Forest Under Threat (Southwest Madagascar): How Public Policy Leads to Conflicting Territories." *Field Actions Science Reports* 3: 1–12.

Blanc-Pamard, Chantal, and Hervé Rakoto Ramiarantsoa. 2003. "Madagascar: Les Enjeux Environnementaux." In *L'Afrique: Vulnérabilité et Défis*, ed. Michael Lesourd, 354–76. Nantes: Editions du Temps.

Blanc-Pamard, Chantal, Florence Pinton, and Hervé Rakoto Ramiarantsoa. 2012. "L'Internationalisation de l'Environnment Madagascar: Un Cas d'Ecole." In *Géopolitique et Environnement: Les Leçons de l'Expérience Malgache*, ed. Hervé Rakoto Ramiarantsoa, Chantal Blanc-Pamard, and Florence Pinton, 13–37. Marseilles, France: IRD Editions.

Borras, Saturnino M., Philip McMichael, and Ian Scoones. 2010. "The Politics of Biofuels, Land and Agrarian Change: Editors' Introduction." *Journal of Peasant Studies* 37(4): 575–92.

Borras, Saturnino M., Ruth Hall, Ian Scoones, Ben White, and Wendy Wolford. 2011. "Towards a Better Understanding of Global Land Grabbing: An Editorial Introduction." *Journal of Peasant Studies* 38(2): 209–16.

Borrini-Feyerabend, Grazia, and Nigel Dudley. 2005a. *Les Aires Protégées à Madagascar: Bâtir le Système à Partir de la Base: Rapport de la Seconde Mission UICN (Version Finale)*. Geneva. World Conservation Union Commission on Environmental, Economic and Social Policy and World Commission on Protected Areas.

———. 2005b. *Elan Durban ... Nouvelles Perspectives Pour les Aires Protégées à Madagascar.* Geneva. World Commission on Protected Areas and Committee on Environmental, Economic and Social Policy of the World Conservation Union, and MIARO.

Bowles, Ian, and Cyril Kormos. 1995. "Environmental Reform at the World Bank: The Role of the U.S. Congress." *Virginia Journal of International Law* 35(4): 777–839.

Bramble, Barbara J., and Gareth Porter. 1992. "Non-Governmental Organizations and the Making of U.S. International Environmental Policy." In *The International Politics of the Environment: Actors, Interests, and Institutions*, ed. Andrew Hurrell and Benedict Kingsbury, 313–53. New York: Oxford University Press.

Brandon, Katrina Eadie, and Michael Wells. 1992. "Planning for People and Parks: Design Dilemmas." *World Development* 20 (4): 557–70.

Bratton, Michael. 1990. "Academic Analysis and U.S. Economic Assistance Policy on Africa." *Issue: A Journal of Opinion* 19(1): 21–37.

Brenner, Neil, and Nik Theodore. 2002. "Cities and the Geographies of 'Actually Existing Neoliberalism.'" *Antipode* 34(3): 349–79.

Brimont, Laura, and Cécile Bidaud. 2014. "Incentivising Forest Conservation: Payments for Environmental Services and Reducing Carbon Emissions from Deforestation." In *Conservation and Environmental Management in Madagascar,* ed. Ivan R. Scales, 299–319. Oxon, England: Earthscan.

Brinkerhoff, Derick W. 1996. "Coordination Issues in Policy Implementation Networks: An Illustration from Madagascar's Environmental Action Plan." *World Development* 24(9): 1497–1510.

Brinkerhoff, Derick W., and Jo Anne Yeager. 1993. *Madagascar's Environmental Action Plan: A Policy Implementation Perspective, Implementing Policy Change Project*. Washington, D.C.: Management Systems International, Abt Associates, Inc., Development Alternatives, Inc., AID / ARTS / FARA's PARTS Project (Policy Analysis, Research and Technical Support).

Brockington, Dan, Jim Igoe, and Kai Schmidt-Soltau. 2006. "Conservation, Human Rights, and Poverty Reduction." *Conservation Biology* 20(1): 250–52.

Brockington, Dan, Rosaleen Duffy, and Jim Igoe. 2008. *Nature Unbound: The Past, Present, and Future of Protected Areas.* London: Earthscan.

Brooks, Thomas M., Mohamed I. Bakarr, Tim Boucher, Gustavo A. B. da Fonseca, Craig Hilton-Taylor, Jonathan M. Hoekstra, Tom Moritz, Silvio Olivieri, Jeff Parrish, Robert L. Pressey, Ana S. L. Rodrigues, Wes Sechrest, Ali Stattersfield, Wendy Strahm, and Simon N. Stuart. 2004. "Coverage Provided by the Global Protected-Area System: Is It Enough?" *BioScience* 54(12): 1081–91.

Brosius, J. Peter. 2006. "Seeing Communities: Technologies of Visualization in Conservation." In *Reconsidering Community: The Unintended Consequences of an Intellectual Romance,* ed. Gerald Creed, 1–34. Santa Fe: School of American Research Press.

Brosius, J. Peter, and Diane Russell. 2003. "Conservation from Above: An Anthropological Perspective on Transboundary Protected Areas and Ecoregional Planning." *Journal of Sustainable Forestry* 17(1 / 2): 39–65.

Brown, Mervyn. 2000. *A History of Madagascar.* Princeton: Markus Wiener.

BSP. 2001. *Biodiversity Support Program Final Report.* Washington, D.C.: USAID. Submitted to the Global Bureau, Environment Center, Office of Environment and Natural Resources of the United States Agency for International Development.

Burawoy, Michael. 1998. "The Extended Case Method." *Sociological Theory* 16(1): 4–33.

Burpee, Gaye, Paige Harrigan, and Tom Remington. 2000. *A Cooperating Sponsor's Field Guide to USAID Environmental Compliance Procedures.* Baltimore: Catholic Relief Services, Program Quality and Support Department.

Burren, Chris. 2007. *Note Conceptuelle: Les Sites KoloAla (SKA) Pour une Gestion Forestière Durable. Manuscript.*

Büscher, Bram. 2013. *Transforming the Frontier: Peace Parks and the Politics of Neoliberal Conservation in Southern Africa.* Durham: Duke University Press.

Büscher, Bram, and Webster Whande. 2007. "Whims of the Winds of Time? Emerging Trends in Biodiversity Conservation and Protected Area Management." *Conservation and Society* 5(1): 22–43.

Büscher, Bram, Sian Sullivan, Katja Neves, Jim Igoe, and Dan Brockington. 2012. "Towards a Synthesized Critique of Neoliberal Biodiversity Conservation." *Capitalism, Nature, Socialism* 23(2): 4–30.

Büscher, Bram, and Robert Fletcher. 2015. "Accumulation by Conservation." *New Political Economy* 20(2): 273–98.

Butterfield, Samuel Hale. 2004. *U.S. Development Aid: An Historic First.* Westport, Conn.: Praeger.

CABS. 2007. "Regional Habitat Monitoring: Madagascar." Accessed August 18, 2007. http://science.conservation.org/portal/server.pt?open=512&obj

ID=755&&PageID=127564&mode=2&in_hi_userid=124186& cached=true.

Campbell, Faith Thompson. 1977. Letter to Mr. Jerry Christiansen, Office of Senator Claiborne Pell. Washington, D.C.: Natural Resources Defense Council.

Campbell, Gwyn. 1993. "The Structure of Trade in Madagascar, 1750–1810." *International Journal of African Historical Studies* 26(1): 111–48.

———. 2005. *An Economic History of Imperial Madagascar, 1750–1895: The Rise and Fall of an Island Empire.* New York: Cambridge University Press.

———. 2013. "Forest Depletion in Imperial Madagascar, c. 1790–1861." In *Contest for Land in Madagascar: Environment, Ancestors, and Development,* ed. Sandra J. T. M. Evers, Gywn Campbell, and Michael Lambek, 63–95. Leiden: Brill.

Campbell, Lisa M., Shannon Hagerman, and Noella J. Gray. 2014. "Producing Targets for Conservation: Science and Politics at the Tenth Conference of the Parties to the Convention on Biological Diversity." *Global Environmental Politics* 14(3): 41–63.

Canavesio, Remy. 2014. "Formal Mining Investments and Artisanal Mining in Southern Madagascar: Effects of Spontaneous Reactions and Adjustment Policies on Poverty Alleviation." *Land Use Policy* 36: 145–54.

Cardiff, Scott G., and Anjara Andriamanalina. 2007. "Contested Spatial Coincidence of Conservation and Mining Efforts in Madagascar." *Madagascar Conservation and Development* 2(1): 28–34.

Carret, Jean-Christophe. 2003. "Comment Financer Durablement le Réseau d'Aires Protégées Terrestres à Madagascar? Rapport de l'Analyse Economique." World Parks Congress Workshop, "Building Comprehensive Protected Areas Systems," Durban, South Africa.

Carrier, James G., and Paige West. 2009. *Virtualism, Governance and Practice: Vision and Execution in Environmental Conservation.* New York: Berghahn Books.

CAS, CI, DWCT, EAZA, ICTE, MBG, MFG, The Field Museum, Claire Kremen, Dean Keith Gilless, Robert Douglas Stone, WASA, WCS, WWF, and Zoo Zürich. 2009. "Malagasy Government's Decree for Precious Wood Export Will Unleash Further Environmental Pillaging." Last Modified October 9, 2009. Accessed October 22, 2014. http://www.wcs.org/press/press-releases/wood-boycott-madagascar.aspx.

Castree, Noel. 2006. "From Neoliberalism to Neoliberalisation: Consolations, Confusions, and Necessary Illusions." *Environment and Planning A* 38: 1–6.

———. 2008. "Neoliberalising Nature: The Logics of Deregulation and Reregulation." *Environment and Planning A* 40(1): 131–52.

———. 2010. "Neoliberalism and the Biophysical Environment 1: What 'Neoliberalism' Is, and What Difference Nature Makes to It." *Geography Compass* 4(12): 1725–33.

———. 2014. "The Capitalist Mode of Conservation, Neoliberalism and the Ecology of Value." *New Proposals: Journal of Marxism and Interdisciplinary Inquiry* 7(1): 16–37.

CBD Secretariat. 2007. "The Convention on Biological Diversity: 2010 Biodiversity Target: 1) Goals and Subtargets and 2) Indicators." Accessed August 18, 2007. http://www.cbd.int/2010-target/goals-targets.aspx.

Center for Public Integrity. 2008. "Lobby Watch: U.S. Agency for International Development." Accessed December 14, 2008. http://projects.publicintegrity.org/lobby/profile.aspx?act=agencies&year=2003&ag=196&sub=1.

Chape, Stuart, Jeremy Harrison, Mark Spalding, and I. Lysenko. 2005. "Measuring the Extent and Effectiveness of Protected Areas as an Indicator for Meeting Global Biodiversity Targets." *Philosophical Transactions: Biological Sciences* 360(1454): 443–55.

Chapin, Mac. 2004. "A Challenge to Conservationists." *World Watch* 17(6): 17–30.

Chatterjee, Pratap, and Matthias Finger. 1994. *The Earth Brokers: Power, Politics and World Development.* New York: Routledge.

Chemonics International. 2004. *Landscape Development Interventions: Final Report Overview: July 1998–December 2003.* Submitted to USAID / Madagascar.

Chhotray, Vasudha, and David Hulme. 2009. "Contrasting Visions for Aid and Governance in the 21st Century: The White House Millennium Challenge Account and DFID's Drivers of Change." *World Development* 37(1): 36–49.

Christiansen, Karin, and Ingie Hovland. 2003. *The PRSP Initiative: Multilateral Policy Change and the Role of Research.* Working Paper 216. London: Overseas Development Institute.

CI. 2003. "Madagascar to Triple Areas Under Protection: Plan Calls for the Creation of a 6-Million-Hectare Network of Terrestrial and Marine Reserves." Last Modified September 16, 2003. Accessed January 14, 2011. http://www.conservation.org/newsroom/pressreleases/Pages/091603_mad.aspx.

———. 2006. *Biodiversity Corridor Planning and Implementation Program (Corridor); Annual Implementation Plan FY 2006: October 1, 2005–September 30, 2006.* Cooperative Agreement No. LAG-A-00-99-00046-00.

———. 2008. "Corporate Partners." Accessed April 4, 2008. http://dev2.conservation.org/discover/partnership/corporate/Pages/default.aspx.

Commission SAPM. 2006. *Procédure de Création des Aires Protégées du Système d'Aires Protégées de Madagascar (SAPM).* Antananarivo, Madagascar. Draft.

Conable, Barber. 1987. "Address to the World Resources Institute." Speech by the President of the World Bank and International Finance Corporation, Washington, D.C., May 5.

Convention for Collaboration with Respect to Endangered Malagasy Fauna Between the Representatives of the Government of Madagascar and the

International Participants. 1987. In Alison Jolly Archives, Division of Rare and Manuscript Collections, Cornell University Library. Available at http://rmc.library.cornell.edu/.

Convention Relative to the Preservation of Fauna and Flora in Their Natural State. Adopted by the London International Conference, 1933.

Cooper, Frederick, and Randall Packard. 1997. *International Development and the Social Sciences: Essays on the History and Politics of Knowledge*. Berkeley: University of California Press.

Coopération Suisse, USAID, NORAD, and BMZ. 1989. Aide Mémoire from the Bilateral Donors, Revue de Aide Mémoire du Mission de l'Evaluation Plan d'Action Environnemental, June 20. In Alison Jolly Archives, Division of Rare and Manuscript Collections, Cornell University Library. Available at http://rmc.library.cornell.edu/.

Corneille, Faith, and Jeremy Shiffman. 2004. "Scaling-up Participation at USAID." *Public Administration and Development* 24(3): 255–62.

Corson, Catherine. 2010. "Shifting Environmental Governance in a Neoliberal World: US AID for Conservation." *Antipode* 42(3): 576–602.

———. 2011. "Territorialization, Enclosure and Neoliberalism: Non-State Influence in Struggles over Madagascar's Forests." *Journal of Peasant Studies* 38(4): 703–26.

———. 2012. "From Rhetoric to Practice: How High-Profile Politics Impeded Community Consultation in Madagascar's New Protected Areas." *Society and Natural Resources* 25(4): 336–51.

Corson, Catherine, and Kenneth I. MacDonald. 2012. "Enclosing the Global Commons: The Convention on Biological Diversity and Green Grabbing." *Journal of Peasant Studies* 39(2): 263–83.

Corson, Catherine, Kenneth I. MacDonald, and Benjamin Neimark. 2013. "Grabbing 'Green': Markets, Environmental Governance and the Materialization of Natural Capital." *Human Geography* 6(1): 1–15.

Corson, Catherine, Rebecca L. Gruby, Rebecca Witter, Shannon Hagerman, Daniel Suarez, Shannon Greenberg, Maggie Bourque, Noella J. Gray, and Lisa M. Campbell. 2014. "Everyone's Solution? Defining and Re-Defining Protected Areas in the Convention on Biological Diversity." *Conservation and Society* 12(2): 71–83.

Corson, Catherine, Kenneth I. MacDonald, and Lisa Campbell. 2014. "Capturing the Personal in Politics: Ethnographies of Global Environmental Governance." *Global Environmental Politics* 14(3): 21–40.

Corson, Catherine, Sarah Brady, Ahdi Zuber, Julianna Lord, and Angela Kim. 2015. "The Right to Resist: Disciplining Civil Society at Rio+20." *Journal of Peasant Studies* 42(3–4): 859–878.

Cowles, Paul, Soava Rakotoarisoa, Haingolalao Rasolonirinamanana, and Vololona Rasoaromanana. 2001. "Facilitation, Participation, and Learning in an Ecoregion-Based Planning Process: The Case of Ageras in Toliara, Madagascar." In *Biological Diversity: Balancing Interests Through Adaptive*

Collaborative Management, ed. L. E. Buck, C. G. Beisler, J. Schelhas, and E. Wollenberg, 407–22. New York: CRC Press.

Craig, David, and Doug Porter. 2003. "Poverty Reduction Strategy Papers: A New Convergence." *World Development* 31(1): 53–69.

Crewe, Emma, and Elizabeth Harrison. 1998. *Whose Development? An Ethnography of Aid*. London: Zed Books.

CRS. 2006. *Earmarks in the FY 2006 Appropriations Acts*. Washington, D.C.: Congressional Research Service.

da Fonseca, Gustavo A. B., Aaron Bruner, Russell A. Mittermeier, Keith Alger, Claude Gascon, and Richard E. Rice. 2005. "On Defying Nature's End: The Case for Landscape-Scale Conservation." *George Wright Forum* 22(1): 46–60.

Débois, Robin. 2009. "Madagascar: Trafic de Bois Précieux (Palissandre et Ebène): La Fièvre de l'Or Rouge Saigne la Forêt Malgache." *Univers Maoré* 13: 8–15.

Décret portant Création de Réserves Naturelles à Madagascar. 1928. Journal Officiel de Madagascar et Dépendances.

Democratic Republic of Madagascar. 1990. Charter of the Malagasy Environment (English Version) Law No. 90–033 of December 21, 1990. In Alison Jolly Archives, Division of Rare and Manuscript Collections, Cornell University Library. Available at http://rmc.library.cornell.edu/.

Dewar, Robert E., and Henry T. Wright. 1993. "The Cultural History of Madagascar." *Journal of World Prehistory* 7(4): 417–66.

Dewar, Robert E., and Alison F. Richard. 2012. "Madagascar: A History of Arrivals, What Happened, and Will Happen Next." *Annual Review of Anthropology* 41(1): 495–517.

Dowie, Mark. 2005. "Conservation Refugees: When Protecting Nature Means Kicking People Out." *Orion* 16–27.

Dudley, Nigel, and Adrian Phillips. 2006. *Forests and Protected Areas: Guidance on the Use of the IUCN Protected Area Management Categories*. Adrian Phillips, Best Practice Protected Area Guidelines Series No. 12. World Commission on Protected Areas (WCPA).

Duffy, Rosaleen. 2005. "Global Environmental Governance and the Challenge of Shadow States: The Impact of Illicit Sapphire Mining in Madagascar." *Development and Change* 36(5): 825–43.

———. 2006. "Non-Governmental Organisations and Governance States: The Impact of Transnational Environmental Management Networks in Madagascar." *Environmental Politics* 15(5): 731–49.

———. 2007. "Gemstone Mining in Madagascar: Transnational Networks, Criminalisation and Global Integration." *Journal of Modern African Studies* 45(2): 185–206.

Durrell, Lee. 1983. Summary of Meeting on Research and Conservation in Madagascar, 1 and 2 February 1983, Written 27 February 1983. In Alison

Jolly Archives, Division of Rare and Manuscript Collections, Cornell University Library. Available at http://rmc.library.cornell.edu/. Jersey, British Isles.

———. 1984. Biological Research by Foreigners in Madagascar, Considered by the International Advisory Group of Scientists (IAGS) and the Malagasy Authorities, 1 March 1983–18 April 1984, Prepared by Lee Durrell, Chairman of IAGS, for Jean-Jacques Petter, Member of IAGS and Liaison between IAGS and WWF-International to Present to WWF-International, Written 13 April 1984. In Alison Jolly Archives, Division of Rare and Manuscript Collections, Cornell University Library. Available at http://rmc.library.cornell.edu/.

———. 1985. Memo to Researchers Who Have Worked or Plan to Work in Madagascar Through IAGS Regarding Meetings in Madagascar 28 October–10 November 1985, Written 20 March 1985. In Alison Jolly Archives, Division of Rare and Manuscript Collections, Cornell University Library. Available at http://rmc.library.cornell.edu/.

Eckholm, Erik. 1975. "The Other Energy Crisis: Firewood." Worldwatch Paper #1. Washington, D.C.: World Watch Institute.

———. 1976. *Losing Ground: Environmental Stress and World Food Prospects.* New York: Norton (for the Worldwatch Institute).

Edwards, Michael, and David Hulme. 1996. "Too Close for Comfort? The Impact of Official Aid on Nongovernmental Organizations." *World Development* 24(6): 961–73.

Epstein, Susan B. 2003. *CRS Report for Congress Foreign Relations Authorization, FY 2004 and FY 2005: State Department, the Millennium Challenge Account, and Foreign Assistance.* Congressional Research Service: Library of Congress. Washington, D.C.

Erdmann, Thomas K. 2008. *Ecoregional Conservation and Development in Madagascar.* Prepared for the 2008 USAID-Madagascar Stock Taking Exercise.

———. 2010. "Eco-Regional Conservation and Development in Madagascar: A Review of USAID-Funded Efforts in Two Priority Landscapes." *Development in Practice* 20(3): 380–94.

Erdmann, Thomas K., and Georges Rakotodrabe. 2008. *A Critical Shortage of Human Resources for Rural Development: Ecoregional Conservation and Development in Madagascar.* ERI Toamasina. Prepared for the 2008 USAID-Madagascar Stock Taking Exercise.

Escobar, Arturo. 1995. *Encountering Development: The Making and Unmaking of the Third World.* Princeton: Princeton University Press.

Esman, Milton. 2003. *Carrots, Sticks, and Ethnic Conflict: Rethinking Development Assistance.* Ann Arbor: University of Michigan Press.

Essex, Jamey. 2013. *Development, Security, and Aid: Geopolitics and Geoeconomics at the U.S. Agency for International Development.* Athens: University of Georgia Press.

Evers, Sandra J. T. M., Perrine Burnod, Rivo Andrianirina Ratsialonana, and André Teyssier. 2011. "Foreign Land Acquisitions in Madagascar: Competing Jurisdictions of Access." In *African Engagements: Africa Negotiating an Emerging Multipolar World*, ed. Ton Diez, Kjell Havnevik, Mayke Kaag, and Terje Oestigaard, 110–32. Leiden: Brill.

Evers, Sandra J. T. M., Gywn Campbell, and Michael Lambek. 2013. "Land Competition and Human–Environment Relations in Madagascar." In *Contest for Land in Madagascar: Environment, Ancestors and Development*, ed. Sandra J. T. M. Evers, Gywn Campbell, and Michael Lambek, 1–19. Leiden: Brill.

Fairfax, Sally K., and Louise Fortmann. 1990. "American Forestry Professionalism in the Third World: Some Preliminary Observations." *Population and Environment: A Journal of Interdisciplinary Studies* 11(4): 259–72.

Fairhead, James, and Melissa Leach. 1996. *Misreading the African Landscape: Society and Ecology in a Forest Savanna Mosaic.* Cambridge: Cambridge University Press.

———. 2003. *Science, Society and Power.* Cambridge: Cambridge University Press.

Fairhead, James, Melissa Leach, and Ian Scoones. 2012. "Green Grabbing: A New Appropriation of Nature?" *Journal of Peasant Studies* 39(2): 237–61.

Falloux, Francois, and Lee Talbot. 1993. *Crisis and Opportunity.* London: Earthscan.

Feeley-Harnik, Gillian. 1995. "Plants and People, Children or Wealth: Shifting Grounds of 'Choice' in Madagascar." *PoLAR* 18(2): 45–64.

———. 2001. "Ravenala Madagascariensis Sonnerat: The Historical Ecology of a 'Flagship Species' in Madagascar." *Ethnohistory* 48(1–2): 31–86.

Fenn, M. 2003. "Learning Conservation Strategies: A Case Study of the Parc National d'Andohahela." In *The Natural History of Madagascar*, ed. Steven Goodman and Jonathan Benstead, 1494–1500. Chicago: University of Chicago Press.

Ferguson, Barry. 2009. "REDD Comes into Fashion in Madagascar." *Madagascar Conservation and Development* 4(2): 132–37.

Ferguson, Barry, Charlie J. Gardner, Mijasoa M. Andriamarovolonona, Tim Healy, Frank Muttenzer, Shirley M. Smith, Neal Hockley, and Mathilde Gingembre. 2014. "Governing Ancestral Land in Madagascar: Have Policy Reforms Contributed to Social Justice?" In *Governance for Justice and Environmental Sustainability: Lessons Across Natural Resource Sectors in Sub-Saharan Africa*, ed. Merle Sowman and Rachel Wynberg, 63–93. Oxon, England: Routledge.

Ferguson, James. 1994. *The Anti-Politics Machine: "Development," Depoliticization, and Bureaucratic Power in Lesotho.* Minneapolis: University of Minnesota Press.

Fisher, William F. 1997. "Doing Good? The Politics and Antipolitics of NGO Practices." *Annual Review of Anthropology* 26(1): 439–64.

Fleck, Robert, and Christopher Kilby. 2001. "Foreign Aid and Domestic Politics: Voting in Congress and the Allocation of USAID Contracts Across Congressional Districts." *Southern Economic Journal* 67(3): 598–617.

Fortmann, Louise. 1995. "Talking Claims: Discursive Strategies in Contesting Property." *World Development* 33(6): 1053–63.

French, Jerome, Greg Booth, John Lichte, Jean-Marc Andriamananatena, A. Fidy, and Orose Venance. 1995. *Final Evaluation of the WWF Debt-for-Nature Project Providing Institutional Support for the Department of Eaux et Forêts.* Washington, D.C. Prepared by the Development Strategies for Fragile Lands (DESFIL) Project, Washington, D.C., for USAID / Madagascar.

Freudenberger, Karen. 2010. *Paradise Lost? Lessons from Twenty-Five Years of USAID Environment Programs in Madagascar.* Contracted for USAID. Washington, D.C. International Resources Group.

Freudenberger, Mark S. 2008. *Ecoregional Conservation and the Ranomafana–Andringitra Forest Corridor: A Retrospective Interpretation of Achievements, Missed Opportunities, and Challenges for the Future.* Antananarivo, Madagascar. Prepared for the 2008 USAID-Madagascar Stock Taking Exercise.

Froger, Géraldine, and Fano Andriamahefazafy. 2003. "Les Stratégies Environnementales des Organisations Internationales dans les Pays en Développement: Continuité ou Ruptures?" *Mondes en Développement* 31(124): 49–76.

Gade, D. W., and A. N. Perkins-Belgram. 1986. "Woodfuels, Reforestation, and Ecodevelopment in Highland Madagascar." *GeoJournal* 12(4): 365–74.

Ganzhorn, Jörg U. 2011. "Conservation through Payments for an Ecosystem Service?" *Madagascar Conservation and Development* 6(2): 55–56.

GAO. 2002. *USAID Relies Heavily on Nongovernmental Organizations, but Better Data Needed to Evaluate Approaches.* Report to the Chairman, Subcommittee on National Security, Veterans Affairs, and International Relations, Committee on Government Reform, House of Representatives. Washington, D.C., General Accounting Office.

———. 2003. *Strategic Workforce Planning Can Help USAID Address Current and Future Challenges.* Foreign Assistance: Report to Congressional Requesters. Washington, D.C., General Accounting Office.

Gardner, Charlie J. 2011. "IUCN Management Categories Fail to Represent New, Multiple-Use Protected Areas in Madagascar." *Oryx* 45(03): 336–46.

Gaylord, Lisa, and Guy Razafindralambo. 2005. *Madagascar's National Environmental Action Plan. Manuscript.*

Geertz, Clifford. 1973. "Thick Description: Toward an Interpretive Theory of Culture." In *The Interpretation of Cultures: Selected Essays*, 3–30. New York: Basic Books.

———. 1988. *Works and Lives.* Stanford: Stanford University Press.

Gezon, Lisa. 1997. "Institutional Structure and the Effectiveness of Integrated Conservation and Development Projects: Case Study from Madagascar." *Human Organization* 56(4): 462–70.

———. 2000. "The Changing Face of NGOs: Structure and Communitas in Conservation and Development in Madagascar." *Urban Anthropology* 29(2): 181–215.

Glaeser, Andreas. 2005. "An Ontology for the Ethnographic Analysis of Social Processes: Extending the Extended-Case Method." *Social Analysis* 49(3): 16–45.

Global Witness and EIA. 2009. *Investigation into the Illegal Felling, Transport and Export of Precious Wood in Sava Region Madagascar.* Global Witness and the Environmental Investigation Agency (EIA-US) in Cooperation with Madagascar National Parks, the National Environment and Forest Observatory, and the Forest Administration of Madagascar.

Goldman, Michael. 2005. *Imperial Nature: The World Bank and Struggles for Social Justice in the Age of Globalization.* New Haven: Yale University Press.

Gore, Albert. 1993. *From Red Tape to Results: Creating a Government that Works Better and Costs Less: Report of the National Performance Review.* Darby, Penn.: Diane Books.

Gore, Charles. 2000. "The Rise and Fall of the Washington Consensus as a Paradigm for Developing Countries." *World Development* 28(5): 789–804.

Gorenflo, Larry, Catherine Corson, Ken Chomitz, Grady Harper, Miroslav Honzak, and Berk Ozler. 2011. "Exploring the Association Between People and Deforestation in Madagascar." In *Human Population: The Demography and Geography of Homosapiens and Their Implications for Biodiversity,* ed. Rich Cincotta and Larry Gorenflo, 197–221. Berlin: Springer.

Gowan, Peter. 1999. *The Global Gamble: Washington's Faustian Bid for World Dominance.* London: Verso Press.

Gramsci, Antonio. 2010 [1971]. *Selections from the Prison Notebooks.* Translated by Q. Hoare and G. N. Smith, *Classic Perspectives on Political Economy.* New York: International.

Greve, Albert. 1990. Madagascar Environment Program Newsletter, December 1990. In Alison Jolly Archives, Division of Rare and Manuscript Collections, Cornell University Library. Available at http://rmc.library.cornell.edu/.

———. 1991a. Madagascar Environment Program Newsletter, May 1991–Volume 1, No. 3. In Alison Jolly Archives, Division of Rare and Manuscript Collections, Cornell University Library. Available at http://rmc.library.cornell.edu/.

———. 1991b. Madagascar Environment Program Newsletter, February 1991–Volume 1, No. 2. In Alison Jolly Archives, Division of Rare and Manuscript Collections, Cornell University Library. Available at http://rmc.library.cornell.edu/.

————. 1992. Madagascar Environment Program Newsletter, March 1992–Volume 2, No. 2. In Alison Jolly Archives, Division of Rare and Manuscript Collections, Cornell University Library. Available at http://rmc.library.cornell.edu/.

————. 1993. Madagascar Environment Program Newsletter, January 1993–Volume 2, No. 4. In Alison Jolly Archives, Division of Rare and Manuscript Collections, Cornell University Library. Available at http://rmc.library.cornell.edu/.

————. 1994. Madagascar Environment Program Newsletter, April 1994–Volume 3, No. 2. In Alison Jolly Archives, Division of Rare and Manuscript Collections, Cornell University Library. Available at http://rmc.library.cornell.edu/.

Greve, Albert Michael, Julian Lampietti, and Francois Falloux. 1995. *National Environmental Action Plans in Sub-Saharan Africa.* Toward Environmentally Sustainable Development in Sub-Saharan Africa, Post UNCED Series: Building Blocks for Africa 2025, Paper No. 6. Environmentally Sustainable Development Division, Africa Technical Department.

Griveaud, P., and R. Albignac. 1972. "The Problems of Nature Conservation in Madagascar." In *Biogeography and Ecology in Madagascar,* ed. R. Battistini and B. Richard-Vindard, 727–39. The Hague: Springer Netherlands.

GTC. 1981. "Global Tomorrow Coalition Inc." *The Environmentalist 1(2):* 168.

Gupta, Akhil. 1997. "Agrarian Populism in the Development of a Modern Nation (India)." In *International Development and the Social Sciences,* ed. F. Cooper and R. Packard, 320–44. Berkeley: University of California Press.

Gusterson, Hugh. 1997. "Studying up Revisited." *Political and Legal Anthropology Review* 20(1): 114–19.

Hajer, Maarten A. 1995. *The Politics of Environmental Discourse: Ecological Modernization and the Policy Process.* Oxford: Clarendon Press.

Hall, Stuart. 1985. "Signification, Representation, Ideology: Althusser and the Post-Structuralist Debates." *Critical Studies in Mass Communication* 2(2): 91–114.

————. 1996. "Gramsci's Relevance for the Study of Race and Ethnicity." In *Critical Dialogues in Cultural Studies,* ed. David Morley and Kuan-Hsing Chen, 411–41. New York: Routledge.

————. 2002. "Reflections on 'Race,' Articulation and Societies Structured in Dominance." In *Race Critical Theories: Text and Context,* ed. Philomena Essed and David Theo Goldberg, 449–54. Oxford: Blackwell.

Hannah, Lee, Berthe Rakotosamimanana, Jorg Ganzhorn, Russell Mittermeier, Silvio Olivieri, Lata Iyer, Serge Rajaobelina, John Hough, Fanja Andriamialisoa, Ian Bowles, and Georges Tilkin. 1998. "Participatory Planning, Scientific Priorities, and Landscape Conservation in Madagascar." *Environmental Conservation* 25(1): 30–36.

Hannerz, Ulf. 2003. "Being There . . . and There . . . and There! Reflections on Multi-Site Ethnography." *Ethnography* 4(2): 201–16.

Harper, Janice. 2002. *Endangered Species: Health, Illness and Death Among Madagascar's People of the Forest*. Durham, N.C.: Carolina Academic Press.

Harrison, Elizabeth. 2003. "The Monolithic Development Machine?" In *A Moral Critique of Development: In Search of Global Responsibilities*, ed. Ananta Kumar Giri and Philip Quarles van Ufford, 101–17. New York: Routledge.

Hart, Gillian. 2001. "Development Critiques in the 1990s: Culs de Sac and Promising Paths." *Progress in Human Geography* 25(4): 649–58.

———. 2002. *Disabling Globalization: Places of Power in Post-Apartheid South Africa*. Berkeley: University of California Press.

———. 2006a. "Post-Apartheid Developments in Comparative and Historical Perspective." In *The Development Decade?: Economic and Social Change in South Africa, 1994–2004*, ed. Vishnu Padayachee, 13–32. Pretoria: Human Sciences Research Council.

———. 2006b. "Denaturalizing Dispossession: Critical Ethnography in the Age of Resurgent Imperialism." *Antipode* 38(5): 977–1004.

Harvey, David. 2003. *The New Imperialism*. Oxford: Oxford University Press.

———. 2005. *A Brief History of Neoliberalism*. New York: Oxford University Press.

Hawkins, Frank, and Ned Horning. 2001. *Forest Cover Change in USAID and Control Areas: A Preliminary Evaluation and Report to USAID and Partners*. International Resources Group Ltd. Projet d'Appui à la Gestion de l'Environnement. Antananarivo, Madagascar.

Healy, Timothy, and Rivo Randriamanantsoa Ratsimbarison. 1998. "Historical Influences and the Role of Traditional Land Rights in Madagascar: Legality Versus Legitimacy." Conference on Land Tenure in Development Cooperation, Capetown, South Africa.

Hecht, Joy E., Dave Gibson, and Brian App. 2008. *Protecting Hard-Won Ground: USAID Experience and Prospects for Biodiversity Conservation in Africa*. Biodiversity Assessment and Technical Support Program (BATS), EPIQ IQC: EPP-I-00-03-00014-00, Task Order 02. Washington, D.C. Chemonics International Inc. for the U.S. Agency for International Development.

Heijnsbergen, P. van. 1997. *International Legal Protection of Wild Fauna and Flora*. Washington, D.C.: IOS Press.

Heniff, Bill, Jr. 2012. *Overview of the Authorization-Appropriations Process*. Congressional Research Service Report for Congress. Washington, D.C. Congressional Research Service.

Henkels, Diane. 2001–2. "A Close-up of Malagasy Environmental Law." *Vermont Journal of Environmental Law*. 3: 1–18.

Hiar, Corbin. 2013. "The Congressman, the Safari King, and the Woman Who Tried to Look Like a Cat: Inside the Shadowy Nonprofit That Throws Some of D.C.'s Finest Junkets." *Mother Jones*, March 12.

————. 2014. "Enviro Groups: Maybe We Shouldn't Help Fund $47,000 Safaris for Polluter-Friendly Politicians." *Mother Jones*, January 22.

Hills, Alice. 2006. "Trojan Horses? USAID, Counterterrorism and Africa's Police." *Third World Quarterly* 27(4): 629–43.

Hoben, Allan. 1989. "USAID: Organizational and Institutional Issues and Effectiveness." In *Cooperation for International Development: The United States and the Third World in the 1990s*, ed. Robert Berg and David Gordon, 253–78. Boulder: Lynne Reinner.

————. 1995. "Paradigms and Politics: The Cultural Construction of Environmental Policy in Ethiopia." *World Development* 23(6): 1007–21.

————. 1997. "The Role of Development Discourse in the Construction of Environment Policy in Africa." PhD Research Seminar on Concepts and Metaphors: Ideologies, Narratives in Myths in Development Discourse Karrebaeksminde, Denmark, December 1–4.

Hockley, Neal J., and Mijasoa M. Andriamarovololona. 2007. *The Economics of Community Forest Management in Madagascar: Is There a Free Lunch? An Analysis of Transfert de Gestion*. Antananarivo, Madagascar: USAID. Development Alternatives, Inc., for the United States Agency for International Development.

Hoekstra, Jonathan M., Timothy M. Boucher, Taylor H. Ricketts, and Carter Roberts. 2005. "Confronting a Biome Crisis: Global Disparities of Habitat Loss and Protection." *Ecology Letters* 8: 23–29.

Holmes, George. 2011. "Conservation's Friends in High Places: Neoliberalism, Networks, and the Transnational Conservation Elite." *Global Environmental Politics* 11(4): 1–21.

————. 2012. "Biodiversity for Billionaires: Capitalism, Conservation and the Role of Philanthropy in Saving / Selling Nature." *Development and Change* 43(1): 185–203.

Horning, Nadia Rabesahala. 2008a. "Strong Support for Weak Performance: Donor Competition in Madagascar." *African Affairs* 107(428): 405–31.

————. 2008b. "Madagascar's Biodiversity Conservation Challenge: From Local- to National-Level Dynamics." *Journal of Integrative Environmental Sciences* 5(2): 109–28.

————. 2012. "Debunking Three Myths About Madagascar's Deforestation." *Madagascar Conservation and Development* 7(3): 116–19.

Hough, John. 1994. "Institutional Constraints to the Integration of Conservation and Development: A Case Study from Madagascar." *Society and Natural Resources* 7(2): 119–24.

Hufty, Marc, and Frank Muttenzer. 2002. "Devoted Friends: The Implementation of the Convention on Biological Diversity in Madagascar." In *Governing Global Biodiversity*, ed. Philip Le Prestre, 279–309. London: Ashgate.

Humbert, Henri. 1933. "Parcs Nationaux et Réserves Naturelles en Afrique et à Madagascar." *Bulletin de l'Association Française pour l'Avancement des Sciences*, 211–15.

Hutton, Jon, William M. Adams, and James C. Murombedzi. 2005. "Back to the Barriers? Changing Narratives in Biodiversity Conservation." *Forum for Development Studies* 32(2): 341–70.

ICCF. 2007a. *ICCF Partners in Conservation*. Washington, D.C.: International Conservation Caucus Foundation.

———. 2007b. "International Conservation Caucus Foundation Overview: Mission and Goals." Accessed October 4, 2014. http://iccfoundation.us/index.php?option=com_content&view=article&id=51&Itemid=66.

———. 2008. "The International Conservation Caucus Foundation: Congressional Briefing Series." Accessed January 15, 2008. http://www.iccfoundation.us/e-briefings/2004.htm.

IDCA. 1981. *U.S. Agency for International Development Annual Budget Submission: FY 1983, Indian Ocean State (Seychelles, Mauritius and Madagascar)*. Washington, D.C., U.S. Agency for International Development.

———. 1985. *U.S. Agency for International Development Annual Budget Submission: FY 1987, Indian Ocean State (Seychelles, Mauritius and Madagascar)*. Washington, D.C., U.S. Agency for International Development.

Igoe, Jim. 2010. "The Spectacle of Nature in the Global Economy of Appearances: Anthropological Engagements with the Spectacular Mediations of Transnational Conservation." *Critique of Anthropology* 30(4): 375–97.

Igoe, Jim, and Dan Brockington. 2007. "Neoliberal Conservation: A Brief Overview." *Conservation and Society* 5(4): 432–49.

Innes, John L. 2010. "Madagascar Rosewood, Illegal Logging and the Tropical Timber Trade." *Madagascar Conservation and Development* 5(1): 6–10.

IRG. 2002. *Final Report: Environmental Management Support Project (Page) May 24, 1999–July 15, 2002*. Environmental Policy and Institutional Strengthening Indefinite Quantity Contract (EPIQ). Partners: International Resources Group, Winrock International, and Harvard Institute for International Development. Subcontractors: PADCO; Management Systems International; and Development Alternatives, Inc. Collaborating Institutions: Center for Naval Analysis Corporation; Conservation International; KNB Engineering and Applied Sciences, Inc.; Keller-Bliesner Engineering; Resource Management International, Inc.; Tellus Institute; Urban Institute; and World Resources Institute.

IUCN. 1972. *Comptes Rendus de la Conférence Internationale sur la Conservation de la Nature et de Ses Ressources à Madagascar*. Antananarivo, Madagascar. Publié avec l'aide financière de l'Organisation des Nations Unies pour l'Education, la Science et la Culture.

———. 1980. *World Conservation Strategy: Living Resource Conservation for Sustainable Development*. Gland, Switzerland. Prepared by the International Union for the Conservation of Nature and Natural Resources with the advice, cooperation, and financial assistance of UNEP and WWF and in collaboration with FAO and UNESCO.

————. 2004. *Governance of Natural Resources—The Key to a Just World that Values and Conserves Nature?* Briefing Note 7. Commission on Environmental, Economic and Social Policy, World Commission on Protected Areas of the World Conservation Union.

Jarosz, Lucy. 1993. "Defining and Explaining Tropical Deforestation: Shifting Cultivation and Population Growth in Colonial Madagascar (1896–1940)." *Economic Geography* 64(9): 366–79.

Jenkins, Clinton N., and Lucas Joppa. 2009. "Expansion of the Global Terrestrial Protected Area System." *Biological Conservation* 12(10): 2166–74.

Jolly, Alison. 2004. *Lords and Lemurs.* Boston: Houghton Mifflin.

————. 2015. *Thank You Madagascar: The Conservation Diaries of Alison Jolly.* London: Zed Books.

Jolly, Alison, and R. W. Sussman. 2007. "Notes on the History of Ecological Studies of Malagasy Lemurs." In *Lemurs: Ecology and Adaptation,* ed. Lisa Gould and Michelle L. Sauther, 19–39. New York: Springer.

Kaufmann, Jeffery C. 2014. "Contrasting Visions of Nature and Landscapes." In *Conservation and Environmental Management in Madagascar,* ed. Ivan R. Scales, 320–341. Oxon, England: Earthscan.

Keck, Andrew, Narendra P. Sharma, and Feder Gershon. 1994. *Population Growth, Shifting Cultivation and Unsustainable Agricultural Development: A Case Study in Madagascar, Africa Technical Department Series.* Washington, D.C.: World Bank Discussion Papers #234.

Keck, Margaret E., and Kathryn Sikkink. 1998. *Activists Beyond Borders: Advocacy Networks in International Politics.* Ithaca: Cornell University Press.

Keller, Eva. 2008. "The Banana Plant and the Moon: Conservation and the Malagasy Ethos of Life in Masoala, Madagascar." *American Ethnologist* 35(4): 650–64.

Kelly, Alice B. 2011. "Conservation Practice as Primitive Accumulation." *Journal of Peasant Studies* 38(4): 683–701.

Klein, Jorgen. 2004. "Fiddling While Madagascar Burns: Deforestation Discourses and Highland History." *Norsk Geografisk Tidsskrift* 58(1): 11–22.

Kormos, Cyril, Brett Grosko, and Russell A. Mittermeier. 2001. "U.S. Participation in International Environmental Law and Policy." *Georgetown International Environmental Law Review* 13(3): 661–693.

Kraemer, Antonie. 2012. "Whose Forests, Whose Voices? Mining and Community-Based Nature Conservation in Southeastern Madagascar." *Madagascar Conservation and Development* 7 (2S): 87–96.

Kremen, C., A. Cameron, A. Moilanen, S. J. Phillips, C. D. Thomas, H. Beentje, J. Dransfield, B. L. Fisher, F. Glaw, T. C. Good, G. J. Harper, R. J. Hijmans, D. C. Lees, E. Louis Jr., R. A. Nussbaum, C. J. Raxworthy, A. Razafimpahanana, G. E. Schatz, M. Vences, D. R. Vieites, P. C. Wright, and M. L. Zjhra. 2008. "Aligning Conservation Priorities across Taxa in Madagascar with High-Resolution Planning Tools." *Science* 320: 222–26.

Kull, Christian A. 1996. "The Evolution of Conservation Efforts in Madagascar." *International Environmental Affairs* 8(1): 50–86.

———. 2002a. "Empowering Pyromaniacs in Madagascar: Ideology and Legitimacy in Community-Based Natural Resource Management." *Development and Change* 33(1): 57–78.

———. 2002b. "Madagascar Aflame: Landscape Burning as Peasant Protest, Resistance, or a Resource Management Tool?" *Political Geography* 21(7): 927–53.

———. 2004. *Isle of Fire: The Political Ecology of Landscape Burning in Madagascar.* Chicago: University of Chicago Press.

———. 2014. "The Roots, Persistence, and Character of Madagascar's Conservation Boom." In *Conservation and Environmental Management in Madagascar,* ed. Ivan R. Scales, 146–71. Oxon, England: Earthscan.

Kull, Christian A., Alain Bertrand, Nadia Rabeshala Horning, Sandra J. T. M. Evers, Bram Tucker, Genese M. Sodikoff, and Jeffrey C. Kaufmann. 2010. "Interview: Social Science and Conservation in Madagascar." *Madagascar Conservation and Development* 5(2): 117–24.

Kull, Steven. 1995. "What the Public Knows that Washington Doesn't." *Foreign Policy* 101 (Winter 1995–96): 102–15.

Kux, Molly. 1991. "Linking Rural Development with Biological Conservation." In *Biodiversity: Culture, Conservation, and Eco-Development,* ed. Margery Olfield and Janice Alcorn, 297–316. Boulder: Westview Press.

Lancaster, Carol. 1999. *Aid to Africa: So Little Done, So Much to Do.* Chicago: University of Chicago Press.

———. 2007. *Foreign Aid: Diplomacy, Development, Domestic Politics.* Chicago: University of Chicago Press.

Langhammer, Penny F., Mohamed I. Bakarr, Leon A. Bennun, Thomas M. Brooks, Rob P. Clay, Will Darwall, Naamal De Silva, Graham J. Edgar, Güven Eken, Lincoln D.C. Fishpool, Gustavo A. B. da Fonseca, Matthew N. Foster, David H. Knox, Paul Matiku, Elizabeth A. Radford, Ana S. L. Rodrigues, Paul Salaman, Wes Sechrest, and Andrew W. Tordoff. 2007. *Identification and Gap Analysis of Key Biodiversity Areas.* ed. Peter Valentine. Best Practice Protected Area Guidelines. Gland, Switzerland. IUCN.

Larner, Wendy. 2003. "Neoliberalism?" *Environment and Planning D: Society and Space* 21(5): 509–12.

Larson, Bruce A. 1994. "Changing the Economics of Environmental Degradation in Madagascar: Lessons from the National Environmental Action Plan Process." *World Development* 22(5): 671–89.

Larson, Patricia, Mark Freudenberger, and Barbara Wyckoff-Baird. 1998. *WWF Integrated Conservation and Development Projects: Ten Lessons from the Field, 1985–1996.* Washington, D.C.: World Wildlife Fund, Biodiversity Support Program.

Lavauden, Louis. 1934. "Histoire de la Législation et de l'Administration Forestière a Madagascar." *Revue des Eaux et Forêts* VIII Series 32e Year 11, November.

Leach, Melissa, and Robin Mearns. 1996. *The Lie of the Land: Challenging Received Wisdom on the African Environment.* Portsmouth, N.H.: Heinemann.

Leisz, Stephen. 1998. "Madagascar Country Profile." In *Country Profiles of Land Tenure: Africa, 1996,* ed. John Bruce, 223–29. Madison, Wis.: Land Tenure Center.

Leisz, Stephen, Andrea Robles, James Gage, Haingo Rasolofonirinamanana, Hantanirina Pulchérie, Rivo Randriamanantsoa Ratsimbarison, Karen Schoonmaker Freudenberger, and Peter Bloch. 1995. *Land and Natural Resource Tenure and Security in Madagascar.* University of Wisconsin, Madison: Land Tenure Center.

Lemos, Maria C., and Arun Agrawal. 2006. "Environmental Governance." *Annual Review of Environment and Resources* 31(1): 297–325.

Lewis, Amanda. 1999. "The Evolving Process of Swapping Debt for Nature." *Colorado Journal of International Environmental Law and Policy* 10(2): 431–67.

Lewis, David, Anthony Bebbington, Simon Batterbury, Alpa Shah, Elizabeth Olson, M. Shameem Siddiqi, and Sandra Duvall. 2003. "Practice, Power and Meaning: Frameworks for Studying Organizational Culture in Multi-Agency Rural Development Projects." *Journal of International Development* 15(5): 541–57.

Lewis, David, and David Mosse. 2006. *Development Brokers and Translators: The Ethnography of Aid Agencies.* Bloomfield, Conn.: Kumarian Press.

Lewis, David, and Paul Opoku-Mensah. 2006. "Moving Forward Research Agendas on International NGOs: Theory, Agency and Context." *Journal of International Development* 18(5): 665–75.

Li, Tania Murray. 2007. *The Will to Improve: Governmentality, Development, and the Practice of Politics.* Durham: Duke University Press.

Lindemann, Stefan. 2004. *Madagascar Case Study: Analysis of National Strategies for Sustainable Development.* Madagascar Case Study Unedited Working Paper. Winnipeg, Canada. International Institute for Sustainable Development, Stratos Inc., and the Environmental Policy Research Centre of the Freie Universität Berlin.

Lister, Sarah. 2003. "NGO Legitimacy: Technical Issue or Social Construct?" *Critique of Anthropology* 23(2): 175–92.

MacDonald, Christine. 2008. *Green Inc: An Environmental Insider Reveals How a Good Cause Has Gone Bad.* Guilford, Conn.: Lyons Press.

MacDonald, Kenneth I., and Catherine Corson. 2014. "Orchestrating Nature: Ethnographies of Nature™ Inc." In *Nature™ Inc: New Frontiers of Environmental Conservation in the Neoliberal Age,* ed. Bram Buscher, Wolfram Dressler, and Robert Fletcher, 44–65. Tucson: University of Arizona Press.

Madagascar Biodiversity Fund. 2015. "About Us." Accessed July 1, 2015. http://www.fapbm.org/.

Mansfield, Becky. 2004. "Rules of Privatization: Contradictions in Neoliberal Regulation of North Pacific Fisheries." *Annals of the Association of American Geographers* 94(3): 565–84.

Marcus, George E. 1995. "Ethnography in / of the World System: The Emergence of Multi-Sited Ethnography." *Annual Review of Anthropology* 24: 95–117.

———. 1998. *Ethnography Through Thick and Thin.* Princeton: Princeton University Press.

Marcus, Richard. 2001. "Seeing the Forest for the Trees: Integrated Conservation and Development Projects and Local Perceptions of Conservation in Madagascar." *Human Ecology* 29(4): 381–97.

———. 2004. *Political Change in Madagascar: Populist Democracy or Neopatrimonialism by Another Name?* Institute for Security Studies. Pretoria, South Africa.

Marcussen, Henrik Secher. 2003. "National Environmental Planning in the Third World: Sustaining the Myths?" *Journal of Transdisciplinary Environmental Studies* 2(1): 1–15.

Margules, Chris R., and Robert L. Pressey. 2000. "Systematic Conservation Planning." *Nature* 405: 243–53.

Markowitz, Lisa. 2001. "Finding the Field: Notes on the Ethnography of NGOs." *Human Organization* 60(1): 40–46.

Maskell, Jack H. 1998. *Lobbying Regulations on Non-Profit Organizations.* Congressional Research Service Report for Congress.

Massey, Doreen. 1999. "Imagining Globalization: Power-Geometries of Time-Space." In *Global Futures: Migration, Environment, and Globalization,* ed. A. Brah, Mary J. Hickman, and Maírtín Mac an Ghaill, 27–44. London: Macmillan.

Mawdsley, Emma. 2007. "The Millennium Challenge Account: Neo-Liberalism, Poverty and Security." *Review of International Political Economy* 14(3): 487–509.

McAfee, Kathleen. 1999. "Selling Nature to Save It? Biodiversity and Green Developmentalism." *Environment and Planning D: Society and Space* 17(2): 133–54.

MCC. 2007. "Millennium Challenge Corporation Selection Criteria." Accessed December 19, 2007. http://www.mcc.gov/selection/index.php

McCarthy, James, and Scott Prudham. 2004. "Neoliberal Nature and the Nature of Neoliberalism." *Geoforum* 35(3): 275–83.

McConnell, William J., and Christian Kull. 2014. "Deforestation in Madagascar: Debates over the Island's Forest Cover and Challenge of Measuring Forest Change." In *Conservation and Environmental Management in Madagascar,* ed. Ivan R. Scales, 67–104. Oxon, England: Earthscan.

McCoy, K. Lynn, and Hajanirina Razafindrainibe. 1997. *Madagascar's Integrated Conservation and Development Projects: Lessons Learned by Participants, Project Employees, Related Authorities and Community Beneficiaries, Executive Summary*. For the SAVEM Steering Committee for Lessons Learned.

MDS. 1990. Madagascar Environment Program I Multidonor Secretariat Mission Report: September 6-October 17, 1990. In Alison Jolly Archives, Division of Rare and Manuscript Collections, Cornell University Library. Available at http://rmc.library.cornell.edu/.

Meadows, Donella H., Dennis L. Meadows, Jorgen Randers, and Williams W. Behrens III. 1972. *Limits to Growth*. A Report for the Club of Rome's Project on the Predicament of Mankind. New York: Universe Books. Potomac Associates.

Medley, Kimberly E. 2004. "Measuring Performance under a Landscape Approach to Biodiversity Conservation: The Case of USAID / Madagascar." *Progress in Development Studies* 4(4): 319–41.

Méral, Phillipe. 2012. "Economie Politique Internationale et Conservation." In *Géopolitique et Environnement: Les Leçons de l'Expérience Malgache*, ed. Hervé Rakoto Ramiarantsoa, Chantal Blanc-Pamard, and Florence Pinton, 73–98. Marseilles, France: IRD Editions.

Méral, Philippe, Géraldine Froger, Fano Andriamahefazafy, and Ando Rabearisoa. 2009. "Le Financement des Aires Protégées à Madagascar: De Nouvelles Modalités." In *Aires Protégées, Espaces Durables?*, ed. Catherine Aubertin and Estienne Rodary, 135–55. Montpellier, France: IRD Editions.

Mercier, Jean-Roger. 2006. "The Preparation of the National Environmental Action Plan (NEAP): Was It a False Start?" *Madagascar Conservation and Development* 1(1): 50–54.

MESupReS, and WWF-International. 1983. Scientific and Cultural Agreement between the Ministry for Higher Education and Scientific Research of the Democratic Republic of Madagascar and the Agency of World Wildlife Fund International, Dated 27 January 1983 with Annexe de l'Accord Scientifique et Culturel Pour la Formation d'un Groupe International Consultatif Chargé de la Coordination Pour Aider à l'Elaboration des Programmes de Conservation à Madagascar, Dated 3 February 1983. In Alison Jolly Archives, Division of Rare and Manuscript Collections, Cornell University Library. Available at http://rmc.library.cornell.edu/.

MFG. 1994. "Editor's Note." *The Madagascar Fauna Group Members' Newsletter*. March.

Milner, Helene V., and Dustin Tingley. 2013. "Public Opinion and Foreign Aid: A Review Essay." *International Interactions* 39(3): 389–401.

Mitchell, Henry. 1987. "On Madagascar, Friends and Lemurs: Wildlife-Watching and Other Pleasures." *Washington Post*, January 25.

Mitchell, Timothy. 2002. *Rule of Experts: Egypt, Techno-Politics, Modernity*. Berkeley: University of California Press.

Mittermeier, Russ. 1986. A Draft Action Plan for Conservation in Madagascar (Revised). In Alison Jolly Archives, Division of Rare and Manuscript Collections, Cornell University Library. Available at http://rmc.library .cornell.edu/.

———. 1987. Letter to Monsieur le Ministre, Ministère de la Production Animale et des Eaux et Forêts (MPAEP), Antananarivo, Madagascar, from Russell Mittermeier, Chairman IUCN / SSC Primate Specialist Group and Vice Chairman, International Programs, IUCN / SSC with Attached Annotated Itinerary for Malagasy Delegation Visiting the United States and Europe (May 10–31, 1987) and Summary of Key Points Discussed During Trip. In Alison Jolly Archives, Division of Rare and Manuscript Collections, Cornell University Library. Available at http://rmc.library.cornell.edu/.

———. 1988. "Primate Diversity and the Tropical Forest: Case Studies from Brazil and Madagascar and the Importance of the Megadiversity Countries." In *Biodiversity*, ed. Edward O. Wilson, 145–54. Washington, D.C.: National Academy Press.

Mittermeier, Russ, and Alison Richard. 1985. Memo to Participants in the SSC Workshop on Species Conservation Priorities in Madagascar, October 30–31, 1985, regarding the Revised Program for the Workshop, Dated October 3, 1985. In Alison Jolly Archives, Division of Rare and Manuscript Collections, Cornell University Library. Available at http:// rmc.library.cornell.edu/.

Mittermeier, Russ, Norman Myers, Jorgen B. Thomsen, Gustavo A. B. da Fonseca, and Silvio Olivieri. 1998. "Biodiversity Hotspots and Major Tropical Wilderness Areas: Approaches to Setting Conservation Priorities." *Conservation Biology* 12(3): 516–20.

Mohan, Giles, and Kristian Stokke. 2000. "Participatory Development and Empowerment: The Dangers of Localism." *Third World Quarterly* 21(2): 247–68.

Montagne, Pierre, and Alain Bertrand. 2006. "Histoire des Politique Forestières au Niger, au Mali et à Madagascar." In *L'Etat et la Gestion Locale Durable des Forêts en Afrique Francophone et à Madagascar,* ed. Alain Bertrand, Pierre Montagne, and Alain Karsenty, 54–83. Paris: Editions Harmattan.

Montagne, Pierre, and Bruno Ramamonjisoa. 2006. "Politiques Forestières à Madagascar entre Répression et Autonomie des Acteurs." *Economie Rurale* 294–95 (July-October): 9–26.

Moore, Donald. 1999. "The Crucible of Cultural Politics: Reworking 'Development' in Zimbabwe's Eastern Highlands." *American Ethnologist* 26(3): 654–89.

———. 2005. *Suffering for Territory: Race, Place, and Power in Zimbabwe.* Durham: Duke University Press.

Moore, Sally Falk. 2001. "The International Production of Authoritative Knowledge: The Case of Drought-Stricken West Africa." *Ethnography* 2(2): 243–71.

Moreau, Sophie. 2008. "Environmental Misunderstandings." In *Greening the Great Red Island: Madagascar in Nature and Culture*, ed. Jeffrey Kaufmann, 58–59. Pretoria: Africa Institute of South Africa.

Mosse, David. 2005. *Cultivating Development: An Ethnography of Aid Policy and Practice*. London: Pluto Press.

Mosse, David, and David Lewis. 2005. *The Aid Effect: Giving and Governing in International Development*. London: Pluto Press.

MRSTD, and Smithsonian Institution. 1988. Memorandum of Understanding between the Ministry of Scientific Research and Technological Development in the Democratic Republic of Madagascar and the Smithsonian Institution in the United States of America to Strengthen Cooperation in Natural Science and Conservation, in Culture and Arts, and in Museum Studies.

Mukonoweshuro, Eliphas G. 1994. "Madagascar: The Collapse of an Experiment." *Journal of Third World Studies* 11(1): 336–68.

Muttenzer, Frank. 2002. "Local Government and the International Biodiversity Regime: Collective Bargaining over State Forests in Madagascar." Ninth Biennial Conference of the International Association for the Study of Common Property: Subtheme, "Globalization, Governance and the Commons," Victoria Falls, Zimbabwe.

Myers, Norman. 1988. "Threatened Biotas: 'Hot Spots' in Tropical Forests." *The Environmentalist* 8(3): 187–208.

———. 1990. "The Biodiversity Challenge: Expanded Hotspots Analysis." *The Environmentalist* 10(4): 243–56.

Myers, Norman, Russell A. Mittermeier, Christina Mittermeier, Gustavo A. B. da Fonseca, and Jennifer Kent. 2000. "Biodiversity Hotspots for Conservation Priorities." *Nature* 403: 853–58.

Nader, Laura. 1972. "Up the Anthropologist: Perspectives Gained from Studying Up." In *Reinventing Anthropology*, ed. D. H. Hymes, 284–311. New York: Pantheon Books.

Neimark, Benjamin. 2010. "Subverting Regulatory Protection of 'Natural Commodities': The Prunus Africana in Madagascar." *Development and Change* 41(5): 929–54.

———. 2012. "Industrializing Nature, Knowledge, and Labour: The Political Economy of Bioprospecting in Madagascar." *Geoforum* 43(5): 980–90.

———. 2013. "The Land of Our Ancestors: Property Rights, Social Resistance, and Alternatives to Land." Land Deals Politics Initiative Working Paper 26, The Land Deal Politics Initiative in collaboration with the Institute for Development Studies and the Initiatives in Critical Agrarian Studies, the International Institute of Social Studies, the Institute for Poverty, Land and Agrarian Studies, and the Polson Institute for Global Development.

Neumann, Rod. 1998. *Imposing Wilderness: Struggles over Livelihood and Nature Preservation in Africa*. Berkeley: University of California Press.

Nicoll, Martin. 1988. WWF-Aires Protégées Project 3746 Bimonthly Report, May, Written 30 May 1988. In Alison Jolly Archives, Division of Rare and Manuscript Collections, Cornell University Library. Available at http://rmc.library.cornell.edu/.

Nicoll, Martin E., and Olivier Langrand. 1989. *Madagascar: Revue de la Conservation et des Aires Protégées.* Gland, Switzerland: World Wide Fund for Nature.

Oates, John F. 1999. *Myth and Reality in the Rain Forest: How Conservation Strategies Are Failing in West Africa.* Berkeley: University of California Press.

Olson, David M., and Eric Dinerstein. 1998. "The Global 200: A Representation Approach to Conserving the Earth's Most Biologically Valuable Ecoregions." *Conservation Biology* 12(3): 502–15.

Olson, David M., Eric Dinerstein, Eric D. Wikramanayake, Neil D. Burgess, George V. N. Powell, Emma C. Underwood, Jennifer A. d' Amico, Illanga Itoua, Holly E. Strand, John C. Morrison, Colby J. Loucks, Thomas F. Allnutt, Taylor H. Ricketts, Yumiko Kura, John F. Lamoreux, Wesley W. Wettengel, Prashant Hedao, and Kenneth R. Kassem. 2001. "Terrestrial Ecoregions of the World: A New Map of Life on Earth." *BioScience* 51(11): 933–38.

Olson, Sherry. 1984. "The Robe of Ancestors: Forests in the History of Madagascar." *Journal of Forest History* 28(4): 174–86.

OSF. 2007. *Rapport sur la Gouvernance Environnementale et Forestière: Domaine 6—Les Domaines Transversaux Nouveaux Textes* n° 02 DTR / ONESF-1s 2007; Analyse Succincte du Champ d'Application et de la Consequence Directe des Nouveaux Textes. Antananarivo, Madagascar.

Ottaway, David B., and Joe Stephens. 2003. "Nation Special Report: The Nature Conservancy." *Washington Post,* May 4–December 21, 2003.

Oxby, Claire. 1985. "Forest Farmers: The Transformation of Land Use and Society in Eastern Madagascar." *Unasylva* 37(148): 42–51.

PACT Inc. 2000. *Sustainable Approaches to Viable Environmental Management (SAVEM), Final Project Technical Report September 26,1991-September 30, 2000.* United States Agency for International Development Cooperative Agreement No. 623–0000-A-00-1035-00 Project No. 687–0110.

PACT Inc, Conservation International, and World Wide Fund for Nature. 2004. *Miray Final Technical Report, USAID Cooperative Agreement No. 687-a-00-98-00150-00.* Washington, D.C. MIRAY Program for Ecoregion-based Conservation and Development.

Paddack, Jean-Paul, and Melissa Moye. 2003. *Madagascar's Experience with Swapping Debt for the Environment: Debt-for-Nature Swaps and Heavily Indebted Poor Country (HIPC) Debt Relief.* Background Paper for the 5th World Parks Congress, Durban, South Africa.

Patlis, Jason M. 2007. Testimony Submitted by Jason M. Patlis, Vice President and Managing Director, U.S. Government Relations, World

Wildlife Fund, on Behalf of World Wildlife Fund, the Nature Conservancy, Conservation International and the Wildlife Conservation Society to the U.S. House of Representatives Appropriations Committee, Subcommittee on State, Foreign Operations, and Related Programs, March 29.

Peck, Jamie. 2004. "Geography and Public Policy: Constructions of Neoliberalism." *Progress in Human Geography* 28(3): 392–405.

Peck, Jamie, and Adam Tickell. 2002. "Neoliberalizing Space." *Antipode* 34(3): 380–404.

Peck, Jamie, and Nik Theodore. 2012. "Follow the Policy: A Distended Case Approach." *Environment and Planning A.* 44(1): 21–30.

Peluso, Nancy. 1992. *Rich Forests, Poor People: Resource Control and Resistance in Java.* Berkeley: University of California Press.

———. 1993. "Coercing Conservation? The Politics of State Resource Control." In *The State and Social Power in Global Environmental Politics,* ed. R. Lipschutz and K. Conca, 46–70. New York: Columbia University Press.

Peluso, Nancy Lee, and Peter Vandergeest. 2001. "Genealogies of the Political Forest and Customary Rights in Indonesia, Malaysia, and Thailand." *Journal of Asian Studies* 60(3): 761–812.

Peluso, Nancy Lee, and Christian Lund. 2011. "New Frontiers of Land Control: Introduction." *Journal of Peasant Studies* 38(4): 667–81.

Perreault, Thomas. 2003. "Changing Places: Transnational Networks, Ethnic Politics, and Community Development in the Ecuadorian Amazon." *Political Geography* 22(1): 61–88.

Peters, Joe. 1998. "Transforming the Integrated Conservation and Development Project (ICDP) Approach: Observations from the Ranomafana National Park Project, Madagascar." *Journal of Agricultural and Environmental Ethics* 11(1): 17–47.

Pollini, Jacques. 2007. "Slash and Burn Cultivation and Deforestation in the Malagasy Rain Forests: Representations and Realities." PhD diss., Cornell University.

———. 2011. "The Difficult Reconciliation of Conservation and Development Objectives: The Case of the Malagasy Environmental Action Plan." *Human Organization* 70(1): 74–87.

Pollini, Jacques, and James P. Lassoie. 2011. "Trapping Farmer Communities Within Global Environmental Regimes: The Case of the GELOSE Legislation in Madagascar." *Society and Natural Resources* 24(8): 814–30.

Porter, David. 1990. *U.S. Economic Foreign Aid: A Case Study of the United States Agency for International Development.* New York: Garland.

Porter, Doug, and David Craig. 2004. "The Third Way and the Third World: Poverty Reduction and Social Inclusion in the Rise of 'Inclusive' Liberalism." *Review of International Political Economy* 11(2): 387–423.

Président de la République. 1962. Décret Instituant le Parc National de l'Isalo (Sous-Préfecture d'Ihosy, Province de Fianarantsoa).

Président du Gouvernement Provisoire. 1958. Décret Instituant le Parc National de la Montagne d'Ambre (District et Province de Diégo Suarez). Ministère de Production.

Raik, Daniela. 2007. "Forest Management in Madagascar: An Historical Overview." *Madagascar Conservation and Development* 2(1): 5–10.

Rajaspera, Bruno, Daniela Raik, Hanta Ravololonanahary, Tom Erdman, Martin Nicoll, Lala Jean Rakotoniaina, Tiana Ramahaleo, Michel Randriambololona, Jean Jacques Jaozandry, Fara Lala Razafy, Paul Raonintsoa, and Patrick Ranjatson. 2008. *Thème "Benefits to Communities."* Prepared for the 2008 USAID-Madagascar Stock Taking Exercise.

Ramanantsoavina, George. 1973. *Note sur la Politique et l'Administration Forestières a Madagascar: Historique—Situation Actuelle.* Tananarive, Madagascar: Service des Eaux et Forêts.

Randriamalala, Hery, and Zhou Liu. 2010. "Rosewood of Madagascar: Between Democracy and Conservation." *Madagascar Conservation and Development* 5(1): 11–22.

Randrianandianina, B. N., L. R. Andriamahaly, F. M. Harisoa, and M. E. Nicoll. 2003. "The Role of Protected Areas in the Management of the Island's Biodiversity." In *The Natural History of Madagascar*, ed. Steven Goodman and Jonathan Benstead, 1423–32. Chicago: University of Chicago Press.

Randrianasolo, Joseph. 1987. Speech of His Excellency, the Minister of Livestock Production, Fisheries, Water, and Forests. St. Catherine's Island, May 13, 1987. In Alison Jolly Archives, Division of Rare and Manuscript Collections, Cornell University Library. Available at http://rmc.library.cornell.edu/.

Rasoavahiny, Laurette, Michèle Andrianarisata, Andriamandimbisoa Razafimpahanana, and Anitry N. Ratsifandrihamanana. 2008. "Conducting an Ecological Gap Analysis for the New Madagascar Protected Area System." Manuscript. Protected Areas Programme.

Razafindralambo, Guy. 2005. *Le Programme Environnemental.* Antananarivo, Madagascar.

Razafitsotra, Jean, Michel Randriambololona, Florent Ravoavy, and Sidonie Rasoarimalala. 2008. *Evolution de l'Approche Ecorégionale à Fianarantsoa.* Prepared for the 2008 USAID-Madagascar Stock Taking Exercise.

Rea, Samuel S. 1998. "Interview as Part of the United States Foreign Assistance Oral History Program, Foreign Affairs Oral History Collection Association for Diplomatic Studies and Training, Interviewed By W. Haven North."

Redclift, Michael R. 1992. "The Meaning of Sustainable Development." *Geoforum* 23(3): 395–403.

Repoblika Demokratika Malagasy. 1984. *Stratégie Malgache Pour la Conservation et le Développement Durable.* Antananarivo, Madagascar. Elaboré par la Commission Nationale de la Stratégie de la Conservation des Ressources Vivantes au Service du Développement National Institué par le Décret Présidentiel n° 84–116 du 4 Avril 1984 et avec l'assistance de l'UICN / WWF.

Repoblikan'i Madagasikara. 1979. Décret n° 79–145 du 14 Juin 1979 Autorisant la Création à Madagascar d'une Représentation du Fonds Mondial Pour la Nature.

———. 1984. Décret n° 84–445 Portant Simultanément Adoption de la Stratégie Malgache Pour la Conservation et le Développement Durable et Création d'une Commission Nationale de Conservation Pour le Développement.

———. 2001. Loi n° 2001 / 05 Portant Code de Gestion des Aires Protégées.

———. 2004a. Arrêté Interministériel Complétant les Dispositions de l'Arrêté n° 7340 / 2004 Portant Création d'un Comité Interministériel des Mines et des Forêts. Ministère de l'Energie et des Mines and Ministère de l'Environnement des Eaux et Forêts. n° 12720 / 2004.

———. 2004b. Arrêté Interministériel Portant Création d' un Comité Interministériel des Mines et des Forêts. Ministère de l'Energie et des Mines and Ministère de l'Environnement des Eaux et Forêts. n° 7340 / 2004.

———. 2004c. Arrêté Interministériel Portant Suspension de l'Octroi de Permis Minier et de Permis Forestier dans les Zones Réservées Comme "Sites de Conservation". Ministère de l'Environnement des Eaux et Forêts and Ministère de l'Energie et des Mines. n° 19560 / 2004.

———. 2005a. Arrêté Portant Prorogation de l'Arrêté n° 20.021 / 2005-MinEnvEF Portant Protection Temporaire de l'Aire Protégée en Création Dénommée « Corridor Forestier Ankeniheny-Zahamena ». Districts d'Ambatondrazaka, Moramanga, Brickaville, Toamasina II, Vavatenina, Région Atsinanana, Alaotra–Mangoro, Analanjirofo, Province Autonome de Toamasina. Ministère de l'Environnement des Eaux et Forêts. n° 379 / 2007.

———. 2005b. Organisant l'Application de la Loi n° 2001-005 du 11 Février 2003 Portant Code de Gestion des Aires Protégées. Décret 2005-013.

———. 2005c. Appliquant l'Article 2 Alinéa 2 de Loi n° 2001 / 15 portant Code des Aires Protégées. Décret 2005–848.

———. 2005d. Portant Modification de Certaines Dispositions de la Loi n° 2001–031 du 08 Octobre 2002 Etablissant un Régime Spécial Pour les Grands Investissements dans le Secteur Minier Malagasy, 25 July. Loi n° 2005-022.

———. 2005e. Portant Modification de Certaines Dispositions de la Loi n° 99–022 du 19 Août 1999 Portant Code Minier, 17 Octobre. Loi n° 2005-021.

———. 2006a. *Madagascar Action Plan 2007–2012: A Bold and Exciting Plan for Rapid Development.* Antananarivo.

———. 2006b. Arrêté Interministériel Portant Protection Temporaire de l'Aire Protégée en Création Dénommée « Corridor Forestier

Fandriana-Vondrozo ». Ministère de l'Energie et des Mines and Ministère de l'Environnement des Eaux et Forêts. n° 16 071 / 2006.

———. 2008a. Loi Portant Refonte du Code de Gestion des Aires Protégées. Présidence de la République. n° 028 / 2008.

———. 2008b. Arrêté Interministériel Portant Mise en Protection Temporaire Globale des Sites Visés par l'Arrêté Interministériel n° 17914 du 18 Octobre 2006 et Levant la Suspension de l'Octroi des Permis Miniers et Forestiers Pour Certains Sites. Ministère de l'Environnement des Forêts et du Tourisme and Ministère de l'Energie et des Mines. n° 18633 / 2008.

———. 2008c. Rapport Final Guide d'Utilisation Durable des Ressources Naturelles dans les Aires Protégées. Ministère de l'Environnement des Eaux et Forêts, Direction Générale des Eaux et Forêts and Commission SAPM.

———. 2009a. Guide de Création des Aires Protégées du Système d'Aires Protégées de Madagascar (SAPM). Ministère de l'Environnement des Eaux et Forêts, Direction Générale des Eaux et Forêts and Commission SAPM.

———. 2009b. Création d'Aires Protégées Mesures de Sauvegarde Cadre de Procédure. Ministère de l'Environnement des Eaux et Forêts, Direction Générale des Eaux et Forêts and Commission SAPM.

———. 2010a. Arrêté Interministériel Portant Création, Organisation et Fonctionnement de la Commission du Système des Aires Protégées de Madagascar. Ministère de l'Environnement des Forêts et du Tourisme and Ministère de l'Energie et des Mines.

———. 2010b. Arrêté Interministériel Modifiant l'Arrêté Interministériel Mine-Forêts n° 18633 du 17 Octobre 2008 Portant Mise en Protection Temporaire Globale des Sites Visés par l'Arrêté Interministériel n° 17914 du 18 Octobre 2006 et Levant la Suspension de l'Octroi des Permis Miniers et Forestiers Pour Certains Sites. Ministère de l'Environnement des Forêts et du Tourisme and Ministère de l'Energie et des Mines. 52005 / 2010.

Republic of Madagascar. 2000. Interim Poverty Reduction Strategy Paper. November 20. Antananarivo.

———. 2003. Poverty Reduction Strategy Paper. Washington, D.C.: International Monetary Fund.

République Démocratique de Madagascar. 1990. Loi n° 90–033 du 21 Décembre 1990 Relative à la Charte de l'Environnement Malagasy. Ministère de l'Economie et du Plan.

Resolve. 2004. *Evaluation et Perspectives des Transferts de Gestion des Ressources Naturelles dans le Cadre du Programme Environnemental 3: Mise en Place d'un Système de Suivi Evaluation du Transfert de Gestion des Ressources Naturelles Renouvelables dans le Cadre du P.E. 3 (Phase 3).* Antananarivo, Madagascar. Consortium Resolve-PCP-IRD.

Rich, Bruce. 1994. *Mortgaging the Earth: The World Bank, Environmental Impoverishment, and the Crisis of Development.* Boston: Beacon Press.

Richard, Alison F., and Robert E. Dewar. 2001. "Politics, Negotiation and Conservation: A View from Madagascar." In *African Rain Forest Ecology and Conservation: An Interdisciplinary Perspective,* ed. William Weber, Lee J. T. White, Amy Vedder and Lisa Naughton-Treves. 535–546. New Haven, CT: Yale University Press.

Richard, Alison F., and Sheila O'Connor. 1997. "Degradation, Transformation, and Conservation." In *Natural Change and Human Impact in Madagascar,* ed. Steven M. Goodman and Bruce D. Patterson, 406–18. Washington, D.C.: Smithsonian Institution Press.

Richard, Alison F., and Joelisoa Ratsirarson. 2013. "Partnership in Practice: Making Conservation Work at Bezà Mahafaly, Southwest Madagascar." *Madagascar Conservation and Development* 8(1): 12–20.

Rodríguez, J. P., A. B. Taber, P. Daszak, R. Sukumar, C. Valladares-Padua, S. Padua, L. F. Aguirre, R. A. Medellín, M. Acosta, A. A. Aguirre, C. Bonacic, P. Bordino, J. Bruschini, D. Buchori, S. González, T. Mathew, M. Méndez, L. Mugica, L. F. Pacheco, A. P. Dobson, and M. Pearl. 2007. "Globalization of Conservation: A View from the South." *Science* 317: 755–56.

Roe, Emery. 1994. *Narrative Policy Analysis: Theory and Practice.* Durham: Duke University Press.

Saboureau, P. 1958. "Considérations sur la Protection de la Nature à Madagascar et à la Côte des Somalis." *Le Naturaliste Malgache* X, 1–2: 135–52.

Sachs, Wolfgang. 1992. *The Development Dictionary: A Guide to Knowledge as Power.* London: Zed Books.

Sahlins, Peter. 1998. *Forest Rites: War of the Demoiselles in Nineteenth-Century France.* 1st ed. Harvard Historical Studies. Cambridge: Harvard University Press.

Sargent, Eva L., and David Anderson. 2003. "The Madagascar Fauna Group." In *The Natural History of Madagascar,* ed. Steven Goodman and Jonathan Benstead, 1543–45. Chicago: University of Chicago Press.

Sarrasin, Bruno. 2001. "Elaboration et Mise en Œuvre du Plan d'Action Environnemental à Madagascar (1987–2001): Construction et Problèmes d'une Politique Publique." PhD diss., Université de Paris I: Panthéon Sorbonne.

———. 2003. "Madagascar: Un Secteur Minier en Emergence, Entre l'Environnement et le Développement." *Afrique Contemporaine* 127–44.

———. 2005. "Environnement, Développement et Tourisme à Madagascar: Quelques Enjeux Politiques." *Loisir et Société / Society and Leisure* 28(1): 163–83.

———. 2006a. "The Mining Industry and the Regulatory Framework in Madagascar: Some Developmental and Environmental Issues." *Journal of Cleaner Production* 14 (3–4): 388–96.

————. 2006b. "Economie Politique du Développement Minier à Madagascar: l'Analyse du Projet QMM à Tolagnaro (Fort-Dauphin)." *VertigO–La Revue en Sciences de l'Environnement* 7(2): 1–14.

————. 2007a. "Le Projet Minier de QIT Madagascar Mineral à Tolagnaro (Fort-Dauphin, Madagascar): Quels Enjeux de Développement?" *Afrique Contemporaine* 221: 205–23.

————. 2007b. "Le Plan d'Action Environnemental Malgache de la Genèse aux Problèmes de Mise en Œuvre: Une Analyse Sociopolitique de l'Environnement." *Revue Tiers Monde* 190: 435–54.

————. 2009. "Mining and Protection of the Environment in Madagascar." In *Mining in Africa: Regulation and Development*, ed. Bonnie Campbell, 150–86. Ottawa: Pluto Press.

Scales, Ivan R. 2012. "Lost in Translation: Conflicting Views of Deforestation, Land Use and Identity in Western Madagascar." *The Geographical Journal*. 178(1): 67–79.

————. 2014a. "The Drivers of Deforestation and the Complexity of Land Use in Madagascar." In *Conservation and Environmental Management in Madagascar*, ed. Ivan R. Scales, 105–25. Oxon, England: Earthscan.

————. 2014b. "The Future of Conservation and Development in Madagascar: Time for a New Paradigm?" *Madagascar Conservation and Development* 9(1): 5–12.

Scherr, Jacob. 1978. "Statement on Behalf of the National Audubon Society, the Natural Resources Defense Council, the Nature Conservancy and the Sierra Club on S. 2646, the International Development Assistance Act of 1978, Regarding Environment, Natural Resources and Development to the Senate Foreign Relations Committee."

Schoonmaker-Freudenberger, Karen. 1999. *Flight to the Forest: A Study of Community and Household Resource Management in the Commune of Ikongo, Madagascar*. Fianarantsoa, Madagascar. Report of a Rapid Rural Appraisal Case Study.

Schuurman, Derek, and Porter P. Lowry. 2009. "The Madagascar Rosewood Massacre." *Madagascar Conservation and Development* 4(2): 98–102.

Scott, James C. 1976. *The Moral Economy of the Peasant: Rebellion and Subsistence in Southeast Asia*. New Haven: Yale University Press.

————. 1998. *Seeing Like a State: How Certain Schemes to Improve the Human Condition Have Failed*. New Haven: Yale University Press.

Seagle, Caroline. 2012. "Inverting the Impacts: Mining, Conservation and Sustainability Claims near the Rio Tinto / QMM Ilmenite Mine in Southeast Madagascar." *Journal of Peasant Studies* 39(2): 447–77.

Seidman, Gay. 2001. " 'Strategic' Challenges to Gender Inequality: The South African Gender Commission." *Ethnography* 2(2): 219–41.

Service de Colonisation. 1913. Décret Etablissant Le Régime Forestier, Applicable à Madagascar. Colonie de Madagascar et Dépendances. 28 Août.

Service des Forêts. 1930. Décret Promulguant dans la Colonie de Madagascar et Dépendances le Décret du 25 Janvier 1930, Réorganisent le Régime Forestier Applicable à Madagascar.

Shaffer, Mark L., Kathryn A. Satterson, Archie Carr, and Simon Stuart. 1987. "The Biological Diversity Program of the U.S. Agency for International Development." *Conservation Biology* 1(4): 280–83.

Shaikh, Asif, Thomas Reardon, Daniel Clay, and Philip DeCosse. 1995. *Dynamic Linkages in Environment and Development in Madagascar.* Washington, D.C., International Resources Group and Michigan State University.

Shaw, Clay, Ed Royce, Tom Udall, and John Tanner. 2005. "Letter from the Co-Chairs of the International Conservation Caucus to Secretary of State Condoleezza Rice and USAID Administrator Andrew Natsios." July 14.

Sikor, Thomas, and Christian Lund. 2009. "Access and Property: A Question of Power and Authority." *Development and Change* 40(1): 1–22.

Simsik, Michael J. 2002. "The Political Ecology of Biodiversity Conservation on the Malagasy Highlands." *GeoJournal* 58(4): 233–42.

———. 2003. "Priorities in Conflict: Livelihood Practices, Environmental Threats, and the Conservation of Biodiversity in Madagascar." Ed.D, Education Policy, Research and Administration, University of Massachusetts.

Sirica, John J. 1975. *Environmental Defense Fund, Inc. et al. (Plaintiffs) v. United States Agency for International Development, et al. (Defendants).* United States District Court for the District of Columbia.

Smillie, Ian. 1997. "NGOs and Development Assistance: A Change in Mind-Set?" *Third World Quarterly* 18(3): 563–77.

———. 1999. "United States." In *Stakeholders: Government–NGO Partnerships for International Development,* eds. Ian Smillie and Henny Helmich, 247–62. London: Earthscan.

Smillie, Ian, Henny Helmich, Tony German, and Judith Randel. 1999. *Stakeholders: Government–NGO Partnerships for International Development.* London: Earthscan.

Smithsonian Institution. 1990. *Annual Report for the Smithsonian Institution for the Year Ended September 30, 1989.* Volume 1988–90. Washington, D.C.: Smithsonian Institution Press.

Smucker, Bob. 1999. *The Nonprofit Lobbying Guide.* Washington, D.C.: Independent Sector.

Sodikoff, Genese. 2005. "Forced and Forest Labor Regimes in Colonial Madagascar, 1926–1936." *Ethnohistory* 52(2): 407–35.

———. 2007. "An Exceptional Strike: A Micro-History of 'People Versus Park' in Madagascar." *Journal of Political Ecology* 14: 10-33.

Steen, Harold K. 1991. *Changing Tropical Forests: Historical Perspectives on Today's Challenges in Central and South America.* Durham: Duke University Press.

Streeter, Sandy. 2004. *The Congressional Appropriations Process: An Introduction.* Congressional Research Service Report for Congress. Order Code 97–684 GOV.

Sullivan, Sian. 2013. "Banking Nature? The Spectacular Financialisation of Environmental Conservation." *Antipode* 45(1): 198–217.

Swanson, Richard A. 1997. *National Parks and Reserves—Madagascar's New Model for Biodiversity Conservation: Lessons Learned Through Integrated Conservation and Development Projects (ICDPs).* United States Agency for International Development, Madagascar Contract No. 623-0110-C-00-1041-00. Tropical Research and Development, Inc.

Tableau Récapitulatif des Contributions des Bailleurs de Fonds au PEIII. 2003.

Takacs, David. 1996. *The Idea of Biodiversity: Philosophies of Paradise.* Baltimore: Johns Hopkins University Press.

Tarrow, Sidney. 2005. *The New Transnational Activism.* Cambridge: Cambridge University Press.

Taylor, Peter J., and Fred H. Buttel. 1992. "How Do We Know We Have Global Environmental Problems? Science and the Globalization of Environmental Discourse." *Geoforum* 23(3): 405–416.

Tegtmeyer, Reiner, Andrea Johnson, Etienne Rasarely, Christian Burren, Doreen Robinson, Martin Bauert, and Jean-Pierre Sorg. 2010. "Interview in Madagascar Conservation and Development: Independent Forest Monitoring Madagascar." *Madagascar Conservation and Development* 5(1): 64–71.

Tendler, Judith. 1975. *Inside Foreign Aid.* Baltimore: Johns Hopkins University Press.

Terbough, John. 1999. *Requiem for Nature.* Washington, D.C.: Island Press.

Thayer, Millie. 2001. "Transnational Feminism: Reading Joan Scott in the Brazilian Sertao." *Ethnography* 2(2): 243–71.

Thrupp, Lori Ann, Susanna B. Hecht, and John Browder. 1997. *The Diversity and Dynamics of Shifting Cultivation: Myths, Realities, and Policy Implications.* Washington, D.C.: World Resources Institute.

Toillier, Aurelie, Sylvie Lardon, and Dominique Hervé. 2008. "An Environmental Governance Support Tool: Community-Based Forest Management Contracts (Madagascar)." *International Journal of Sustainable Development* 11(2–4): 187–205.

TRD. 1997. *Final Report of Activities: November 1991–1996, with Special Focus on 1994–1996.* Submitted by Tropical Research and Development Inc. to United States Agency for International Development, Contract number 623-0110-c-00-1040-00. Submitted by TRD, Gainesville Florida.

Tsing, Anna L. 2005. *Friction: An Ethnography of Global Connection.* Princeton: Princeton University Press.

Tucker, Michael. 1994. "A Proposed Debt-for-Nature Swap in Madagascar and the Larger Problem of LDC Debt." *International Environmental Affairs* 6(1): 59–68.

U.S. Congress. 1973. Foreign Military Sales and Assistance Act. *A Bill to Authorize the Furnishing of Defense Articles and Services to Foreign Countries and International Organizations (Including New Directions Legislation).* Public Law 93–189.

———. 1977. International Development and Food Assistance Act. *Title I, International Development Assistance, of a Bill to Amend the Foreign Assistance Act of 1961 to Authorize Development Assistance Programs for Fiscal Year 1978, to Amend the Agricultural Trade Development and Assistance Act of 1954 to Make Certain Changes in the Authorities of That Act.* Public Law 95–88.

———. 1978. International Development and Food Assistance Act. *Title I, Development Assistance, of a Bill to Amend the Foreign Assistance Act of 1961 to Authorize Development and Economic Assistance Programs for Fiscal Year 1979, to Make Certain Changes in the Authorities of That Act and the Agricultural Trade Development and Assistance Act of 1954, and to Improve the Coordination and Administration of U.S. Development-Related Policies and Programs.* Public Law 95–424.

———. 1981a. International Security and Development Cooperation Act. *Title III, Development Assistance, of an Original Bill to Amend the Foreign Assistance Act of 1961 and the Arms Export Control Act to Authorize Appropriations for Development and Security Assistance Programs for Fiscal Year 1982, to Authorize Appropriations for the Peace Corps for the Fiscal Year 1982, to Provide Authorities for the Overseas Private Investment Corporation, and for Other Purposes.* Public Law 97–113.

———. 1981b. Foreign Assistance Legislation for Fiscal Year 1982. Hearing before the Policy Committee on Foreign Affairs: Subcommittee on International Economic and Trade.

———. 1983. International Environmental Protection Act. *Title VII, International Environmental Protection, of a Bill to Authorize Appropriations for Fiscal Years 1984 and 1985 for the Department of State, the United States Information Agency, the Board for International Broadcasting, the Inter-American Foundation, and the Asia Foundation, to Establish the National Endowment for Democracy, and for Other Purposes.* Public Law 98–164.

———. 1986a. Special Foreign Assistance Act. *Title III, Protecting Tropical Forests and Biological Diversity in Developing Countries, of a Bill to Amend the Foreign Assistance Act of 1961 to Provide Assistance to Promote Immunization and Oral Rehydration, and for Other Purposes.* Public Law 99–529.

———. 1986b. Senate Report 99–443 to Accompany S.2824. *Foreign Assistance and Related Programs Appropriations Act, 1987.*

———. 1995. Senate Report 104–143 to Accompany H.R.1868. *Foreign Operations, Export Financing, and Related Programs Appropriations Act, 1996.* Public Law 104–107.

———. 1999. House Conference Report 106–479 to Accompany H.R. 3194. *Making Consolidated Appropriations for the Fiscal Year Ending September 30, 2000, and for Other Purposes.* Public Law 106–113.

———. 2007. Senate Report 110–128 to Accompany H.R.2764. *Making Appropriations for the Department of State, Foreign Operations, and Related Programs for the Fiscal Year Ending September 30, 2008, and for Other Purposes.* Public Law 110–161.

———. 2008. Senate Report 110–425 to Accompany S.3288. *Making Appropriations for the Department of State, Foreign Operations, and Related Programs for the Fiscal Year Ending September 30, 2009, and for Other Purposes.* Public Law 111–47.

———. 2010. Senate Report 111–237 to Accompany S.3676. *Making Appropriations for the Department of State, Foreign Operations, and Related Programs for the Fiscal Year Ending September 30, 2011, and for Other Purposes.* Public Law 112–10.

U.S. Department of State. 1979. *U.S. Agency for International Development Annual Budget Submission: FY 1981, Indian Ocean States.* Washington, D.C.

———. 2006. *U.S. Foreign Assistance Reform: Achieving Results and Sustainability in Support of Transformational Diplomacy.* Washington, D.C.

———. 2007a. *Foreign Assistance Framework.* Director of U. S. Foreign Assistance. January 29.

———. 2007b. "Supplemental Reference: Foreign Assistance Standardized Program Structure and Definitions." October 15.

U.S. Embassy-Madagascar. 2006. Press Release: USAID, QIT Madagascar Minerals (QMM), and the Region of Anosy Join to Fight Poverty, Environmental Degradation and Poor Health Conditions in Southern Madagascar.

U.S. House of Representatives. 1985. U.S. Policy on Biological Diversity. International Organizations Subcommittee on Human Rights, House Foreign Affairs Committee. Hearing, June 6.

U.S. House of Representatives. 2009. *Resolution Condemning the Illegal Extraction of Madagascar's Natural Resources.* H. Res 839. November 4.

U.S. Department of State. 1980. *The World's Tropical Forests: A Policy, Strategy, and Program for the United States.* Report to the President by a U.S. Interagency Task Force on Tropical Forests.

UN. 1972. *Report of the United Nations Conference on the Human Environment, 5–16 June, Stockholm.* 21st plenary meeting.

———. 1987. *Our Common Future.* Report of the Brundtland Commission, A/42/427, to the UN General Assembly, Development and International Co-operation: Environment.

UN Statistics Division. 2008. "Progress Towards the Millennium Development Goals, 1990–2005 Goal 7–Ensure Environmental Sustainability." Department of Economic and Social Affairs Statistics Division. Accessed April 1, 2008. http://unstats.un.org/unsd/mi/goals_2005/Goal_7_2005.pdf.

USAID. 1979. *Environmental and Natural Resource Management in Developing Countries: A Report to Congress. Volume I.* Submitted in response to section

118 of the Foreign Assistance Act of 1961, 22 U.S.C.§ 2151p, as amended by section 110 of the International Development and Food Assistance Act of 1978, PL 95–424 92 Stat. 948. Washington, D.C.

———. 1981. *Congressional Presentation Fiscal Year 1982*. Washington, D.C.

———. 1983a. *Annual Budget Submission: FY 1985, Indian Ocean States: Madagascar*. Washington, D.C.

———. 1983b. *Policy Determination 6: Environmental and Natural Resource Aspects of Development Assistance*. Washington, D.C.

———. 1983c. *Policy Determination 7: Forestry Policy and Programs*. Washington, D.C.

———. 1984a. *Country Development Strategy Statement, FY 1986: Madagascar, Revised Version*. Washington, D.C.

———. 1984b. *Annual Budget Submission: FY 1986, Indian Ocean States: Madagascar*. Washington, D.C.

———. 1985. *U.S. Strategy on the Conservation of Biological Diversity: An Interagency Task Force Report to Congress*. Washington, D.C.

———. 1987. *Annual Budget Submission: FY 1989, Indian Ocean States: Madagascar*. Washington, D.C.

———. 1988a. *Annual Budget Submission: FY 1990, Indian Ocean States: Madagascar*. Washington, D.C.

———. 1988b. *A Concept Paper on Aid Strategy in Madagascar for 1987–1990*. Washington, D.C.

———. 1988c. *USAID Policy Paper: Environment and Natural Resources*. Program Bureau for and Coordination Policy. Washington, D.C.

———. 1991. *The Environment Initiative Progress Update*. Washington, D.C.

———. 1992a. *Madagascar Country Program Strategic Plan (CPSP) FY 1993–1998*. Antananarivo, Madagascar.

———. 1992b. *Environment Strategy*. Washington, D.C.

———. 1994a. USAID Madagascar Environment Program. Draft Manuscript. In Alison Jolly Archives, Division of Rare and Manuscript Collections, Cornell University Library. Available at http://rmc.library.cornell.edu/.

———. 1994b. *Biodiversity Conservation and Sustainable Use: USAID Project Profiles*. The Environment and Natural Resources Information Center. Environmental Project Profile. Washington, D.C.

———. 1994c. *USAID's Biodiversity Program: Conservation and Sustainable Use of Biodiversity*. Environment and Natural Resources Information Center. Washington, D.C.

———. 1995a. *Core Report of the New Partnerships Initiative*. Washington, D.C.

———. 1995b. *USAID / Madagascar Environment Program*. Booklet. Antananarivo, Madagascar.

———. 1996. *Congressional Budget Justification FY 1997*. Washington, D.C.

———. 1997a. *Congressional Budget Justification FY 1998*. Washington, D.C.

————. 1997b. *Environmental Phase 2, 1997–2002.* Antananarivo, Madagascar.

————. 1997c. *Request for Applications (RFA) for a Cooperative Agreement (CA) Under USAID / Madagascar Natural Strategic Agreement to Conserve Biologically Diverse Ecosystems in Priority Conservation Zones.* Issuance Date: December 4, 1997; Closing Date: March 3, 1998. Antananarivo, Madagascar.

————. 1997d. *Madagascar CSP: Country Strategic Plan, 1998–2002.* Antananarivo, Madagascar.

————. 1998a. *A Guide to the Global Environment Center's People, Programs, Projects and Partners.* Washington, D.C.

————. 1998b. *Madagascar Country Strategic Plan FY 1998–2002 Amendment.* Antananarivo, Madagascar.

————. 2002a. *The Full Measure of Foreign Aid.* Foreign Aid in the National Interest: Promoting Freedom, Security and Opportunity. Washington, D.C.

————. 2002b. *U.S. Agency for International Development Policy Guidance USAID—U.S. PVO Partnership, Mandatory Reference: 200–203.* Washington, D.C., Bureau for Policy and Program Coordination, Bureau for Democracy, Conflict and Humanitarian Assistance, Office of Private and Voluntary Cooperation.

————. 2002c. *Madagascar Integrated Strategic Plan FY 2003–2008, Unrestricted Version, (Revised) November.* Antananarivo, Madagascar.

————. 2003a. *Request for Proposals: Eco-Regional Initiatives to Promote Alternatives to Slash and Burn Practices.* Antananarivo, Madagascar.

————. 2003b. *Request for Applications: Environment / Rural Development Strategic Objective: Program Description for Maintaining Biological Integrity of Critical Biodiversity Habitats.* Antananarivo, Madagascar.

————. 2003c. *Request for Proposals: Support Sustainable Environment and Forest Ecosystems Management in Madagascar.* Antananarivo, Madagascar.

————. 2004a. "USAID Forest Program: Over 25 Years of Experience." Paper submitted to the XII World Forestry Congress. Quebec City, Canada.

————. 2004b. *White Paper: U.S. Foreign Aid: Meeting the Challenges of the Twenty-First Century.* Washington, D.C., Bureau for Policy and Program Coordination, U.S. Agency for International Development.

————. 2006. *U.S. International Food Assistance Report 2006.* Washington, D.C.

————. 2007a. *USAID's Biodiversity Conservation and Forestry Programs, FY 2005.* Washington, D.C., U.S. Agency for International Development.

————. 2007b. "USAID's Definition of Biodiversity Programs." Accessed January 9, 2007. http://www.usaid.gov/our_work/environment/biodiversity/code.html.

————. 2008. *Audit of USAID / Madagascar's Biologically Diverse Forest Ecosystem Conservation Program Audit Report No. 4–687–08–002-P.* General Office of Inspector. Pretoria, South Africa.

————. 2009. *Biodiversity Conservation and Forestry Programs Annual Report.* Washington, D.C.

———. 2013. *Biodiversity Conservation and Forestry Programs, FY 2012 Results and Funding: Conserving Biodiversity, Sustaining Forests.* Washington, D.C., USAID.

———. 2014a. *USAID Biodiversity Policy.* Washington, D.C., USAID.

———. 2014b. "USAID, Where We Work, Africa, Madagascar, Environment." Last Modified September 8, 2014. Accessed September 15, 2014. http://www.usaid.gov/madagascar/environment.

USBM. 1894. *United States Bureau of Manufactures Monthly Consular and Trade Reports; Commerce, Manufactures, Etc.* Washington, D.C.: Government Printing Office.

USDA. 2008. "Food Aid Programs: U.S. Food Aid Programs Descriptions: Public Law 480, Food for Progress and Section 416(B)." Accessed May 1, 2008. http://www.fas.usda.gov/excredits/FoodAid/Title%201/pl480ofst. html.

van den Berg, Rob, and Phillip Quarles van Ufford. 2005. "Disjuncture and Marginality—Towards a New Approach to Development Practice." In *The Aid Effect: Giving and Governing in International Development,* ed. David Mosse and David Lewis, 196–212. London: Pluto Press.

Vandergeest, Peter, and Nancy Peluso. 1995. "Territorialization and State Power in Thailand." *Theory and Society* 24: 385–426.

———. 2006a. "Empires of Forestry: Professional Forestry and State Power in Southeast Asia, Part 2." *Environment and History* 12(4): 359–93.

———. 2006b. "Empires of Forestry: Professional Forestry and State Power in Southeast Asia, Part 1." *Environment and History* 12(1): 31–64.

Vega Media. 2009. Madagascar Broader Horizons. *Business Week Special Advertising Section* January 16. Accessed June 17, 2014. http://www. vegamedia.com/reports/new/reports/madagascar_january2009.pdf

Vincent, Carol Hardy, and Jim Monke. 2009. *Earmarks Disclosed by Congress: FY 2008 and FY 2009 Regular Appropriations Bills.* Washington, D.C.: Congressional Research Service.

Vincent, Robert Montgomery. 1991. "Biological Diversity and Third World Development: A Study of the Transformation of an Ecological Concept into Natural Resource Policy." PhD diss., Oregon State University.

Vokatry ny Ala. 2006. *Evaluation et Analyse du Dynamique de Transferts de Gestion dans le Corridor Ranomafana–Andringitra: Cas d'Ampatsy et d'Andranomiditra.* Une collaboration entre EcoRegional Initiatives (ERI)–Fianarantsoa et l'Université de Californie à Berkeley Beahrs Leadership Program. Fianarantsoa, Madagascar. EcoRegional Initiatives—Fianarantsoa, and the University of California Berkeley Beahrs Leadership Program.

Wade, Robert. 1997. "Greening the Bank: The Struggle over the Environment." In *The World Bank: Its First Half Century,* ed. Devesh Kapur, John Lewis and Richard Webb, 611–734. Washington, D.C.: Brookings Institution.

Waeber, Patrick O. 2009. "Madagascar—'Down the River Without a Paddle' or 'Turning the Corner'?" *Madagascar Conservation and Development* 4(2): 72–74.

———. 2012. "Biodiversity Offsetting—en Vogue in Madagascar?" *Madagascar Conservation and Development* 7(3): 110–11.

Walsh, Andrew. 2005. "The Obvious Aspects of Ecological Underprivilege in Ankarana, Northern Madagascar." *American Anthropologist* (4): 654.

Wang, Yiting, and Catherine Corson. 2014. "The Making of a Pro-Poor Carbon Credit: Clean Cookstoves and 'Uncooperative' Women in Western Kenya." *Environment and Planning A* 47(10): 2064–2079.

Watts, Michael. 2001. "Development Ethnographies." *Ethnography* 2(2): 283–300.

Webb, Tim. 2010. "Madagascar Oil Brings Tar Sands Project to London Market." *Guardian, Online,* November 29. Accessed November 22, 2015. http://www.theguardian.com/business/2010/nov/29/oil-oilandgascompanies.

Wedel, Janine R., Cris Shore, Gregory Feldman, and Stacy Lathrop. 2005. "Toward an Anthropology of Public Policy." *Annals of the American Academy of Political and Social Science* 600(1): 30–51.

Weiss, Thomas G. 2005. "Governance, Good Governance and Global Governance: Conceptual and Actual Challenges." In *The Global Governance Reader,* ed. R. Wilkinson, 68–88. London: Routledge.

Weissman, Stephen. 1995. *A Culture of Deference: Congress's Failure of Leadership in Foreign Policy.* New York: Basic Books.

Wilmé, Lucienne, Derek Schuurman, Porter P. Lowry II, and Peter H. Raven. 2009. "Precious Trees Pay Off—but Who Pays?" *Madagascar Conservation and Development* 4 (2): 98–102. Supplementary material for "The Madagascar Rosewood Massacre."

Wilmé, Lucienne, and Patrick O. Waeber. 2010. "Rumblings in the Forests Heard Loudly on the Web." *Madagascar Conservation and Development* 5(1): 2–3.

Wilshusen, Peter R., Steven R. Brechin, Crystal L. Fortwangler, and Patrick C. West. 2002. "Reinventing a Square Wheel: Critique of a Resurgent 'Protection Paradigm' in International Biodiversity Conservation." *Society and Natural Resources* 15(1): 17–40.

Wilson, Edward O. 1988. *Biodiversity.* Washington, D.C.: National Academy Press.

Wilson, Kenneth. 2005. "Of Diffusion and Context: The Bubbling Up of Community-Based Resource Management in Mozambique in the 1990s." In *Communities and Conservation: Histories and Politics of Community-Based Natural Resource Management,* ed. J. Peter Brosius, Anna Lowenhaupt Tsing, and Charles Zerner. 149–176. New York: AltaMira Press.

Winterbottom, Bob. 2001. *Reflections on Improving the Management of Forest Resources in Madagascar.* Prepared for USAID / Madagascar by International Resources Group, Task Order No. 813 Contract No. PCE-I-00-96-00002-00. Washington, D.C.

Wolmer, William. 2003a. "Transboundary Conservation: The Politics of Ecological Integrity in the Great Limpopo Transfrontier Park." *Journal of Southern African Studies* 29(1): 261–78.

———. 2003b. "Transboundary Protected Area Governance: Tensions and Paradoxes." Transboundary Protected Areas in the Governance Stream of the 5th World Parks Congress, Durban, South Africa, September.

World Bank. 1989. Madagascar Plan d'Action Environnementale (PAE) Project Environnement I Mission d'Evaluation des Bailleurs de Fonds Aide Mémoire (Version Préliminaire 18 Juin). In Alison Jolly Archives, Division of Rare and Manuscript Collections, Cornell University Library. Available at http://rmc.library.cornell.edu/.

———. 1990a. *Staff Appraisal Report Democratic Republic of Madagascar Environment Program.* Agriculture Sector Operations Division South Central Region and Africa Indian Ocean Department. Washington, D.C.: World Bank.

———. 1990b. Aide-Mémoire de la Mission de Supervision du Projet Environnement I (du 23 au 27 Juillet 1990), with Cover Memo to Son Excellence Monsieur Jean Robiarivony, Ministre de l'Economie et du Plan, from José Bronfman, Représentant Résident. In Alison Jolly Archives, Division of Rare and Manuscript Collections, Cornell University Library. Available at http://rmc.library.cornell.edu/.

———. 1996. *Environmental Assessments and National Action Plans.* Operations Evaluation Department, Precis and Briefs. No. 130. Washington, D.C.

———. 1997. *Second Environment Program Support Project, Madagascar.* Project Information Document. Washington, D.C.: World Bank.

———. 2001. "Privatization in Madagascar Country Fact Sheet." MIGA and the Africa Region of the World Bank. Accessed June 24, 2014. http://www.fdi.net/documents/WorldBank/databases/plink/factsheets /madagascar.htm.

———. 2003a. *Madagascar: Rural and Environment Sector Review.* Environmentally and Socially Sustainable Development, Africa Region.

———. 2003b. *Project Appraisal Document on a Proposed Credit in the Amount of 23.2 SDR (US$ 32 Million Equivalent) to the Republic of Madagascar for a Mineral Resources Governance Project.* Report No: 25777. Washington, D.C.: World Bank.

———. 2003c. *Implementation Completion Report (If-N0090) on a Credit in the Amount of US$ 30 Million to Madagascar for the Environment Program Phase II Project (P001537 and P040596).* Environment and Social Development Department, Africa Region. Report No. 29309.

———. 2004a. *Madagascar Third Environment Programme (EP III), Project Brief, Project Number P074235.* Implemented jointly by the United Nations Development Programme and the World Bank via the Ministry of Environment, Water and Forestry. Draft Global Environment Fund application.

———. 2004b. *Project Appraisal Document on a Proposed IDA Grant in the Amount of SDR 26.8 Million (US$40 Million Equivalent) and a Grant from the Global Environment Facility Trust Fund in the Amount of US$9 Million to the Republic of Madagascar for a Third Environment Program Support Project.* Environmentally and Socially Sustainable Development—AFTS1, Country Department 8, Africa Regional Office.

———. 2005. "Environmental and Social Safeguard Policies—Policy Objectives and Operational Principles." In *World Bank Operational Manual: Operational Policies Section 4.12.* Washington, D.C.: World Bank.

———. 2007. *Project Performance Assessment Report Madagascar Environment II (Credit N009).* Washington, D.C. Sector Thematic and Global Evaluation Division Independent Evaluation Group.

———. 2010. *Madagascar—Governance and Development Effectiveness Review: A Political Economy Analysis of Governance in Madagascar.* Madagascar: World Bank.

———. 2011a. *Project Paper on a Proposed Additional IDA Credit in the Amount of SDR26 Million (US$ 42 Million Equivalent) and a Proposed Additional Grant from the Global Environment Facility Trust Fund in the Amount of US$ 10.0 Million to the Republic of Madagascar for the Third Environmental Program Support Project (EP3).*

———. 2011b. *The Changing Wealth of Nations: Measuring Sustainable Development in the New Millennium.* Environment and Development. Washington, D.C. International Bank for Reconstruction and Development.

———. 2014. "Madagascar: World Development Indicators." Accessed February 1, 2014. http://data.worldbank.org/country/madagascar#cp_wdi.

World Bank, USAID, Coopération Suisse, UNESCO, UNDP, and WWF. 1988. *Madagascar Plan d'Action Environnemental: Volume I—Document de Synthèse Générale et Propositions d'Orientations, Version Préliminaire.* Antananarivo, Madagascar.

World Bank, USAID, Cooperation Suisse, UNDP, WWF, and UNESCO. 1988. Madagascar Environmental Action Plan: Volume 1, General Synthesis and Proposed Actions, Preliminary Version.

Wright, Henry T., and Jean-Aimé Rakotoarisoa. 2003. "The Rise of Malagasy Societies: New Developments in the Archaeology of Madagascar." In *The Natural History of Madagascar,* ed. Steven M. Goodman and Jonathan P. Benstead, 112–19. Chicago: University of Chicago Press.

WWF, CI, WCS, MDG, DWCT, ICTE, The Peregrine Fund, Fanamby, MFG, l'Homme et l'Environnement, and PROTA. 2009. "Communiqué: The Trees Must Not Hide the Forest: The Loss of Malagasy Heritage."

Yager, Thomas R. 2009. "The Mineral Industry of Madagascar." In *U.S. Geological Survey Minerals Yearbook 2007 [Advance Release].* Reston, Va.: USGS / U.S. Department of Interior.

Zaidi, S. Akbar. 1999. "NGO Failure and the Need to Bring Back the State." *Journal of International Development* 11(2): 259–71.

Zeller, Shawn. 2004. "On the Work Force Roller Coaster at USAID." *Foreign Service Journal* (April): 33–39.

Zoo Zürich. 2009a. "Communiqué de Presse: Madagascar: Un Décret Qui Empire la Crise de la Biodiversité!" Signed by CAS, CI, DWCT, EAZA, ICTE, MBG, MFG, The Field Museum, Chicago, Dr. Claire Kremen, Dean Keith Gilless, Robert Douglas Stone, WAZA, WCS, WWF, Zoo Zürich. 12 Octobre.

———. 2009b. "Communiqué de Presse: Masoala—l'Oeil de la Forêt est Menacé par des Coupes de Bois Illégales." 31 Août.

Index

Page numbers in *italics* refer to illustrations.

295